GENETICS

GENETICS

Issues of Social Justice

edited by Ted Peters

The Pilgrim Library of Ethics

THE PILGRIM PRESS, *Cleveland, Ohio*

The Pilgrim Press, Cleveland, Ohio 44115
© 1998 by Ted Peters

Grateful acknowledgment is made to the following for permission to reprint previously published material:
  Dialog for permission to reprint, with revisions, Philip Hefner, "Determinism, Freedom, and Moral Failure," vol. 33, no. 1 (winter 1994): 23–29; Karen Lebacqz, "Genetic Privacy: No Deal for the Poor," vol. 33, no. 1 (winter 1994): 39–48.
  CTNS Bulletin for permission to reprint Karen Lebacqz, "Fair Shares: Is the Genome Project Just?" vol. 13, no. 4 (fall 1993): 1–6. The *CTNS Bulletin* is published quarterly by the Center for Theology and the Natural Sciences, an affiliate center of the Graduate Theological Union in Berkeley, California.

Printed in the United States of America on acid-free paper

03 02 01 00 99 98   5 4 3 2 1

Library of Congress Cataloging-in-Publication Data

Genetics : issues of social justice / edited by Ted Peters.
      p.   cm. — (The pilgrim library of ethics)
    Includes bibliographical references and index.
    ISBN 0-8298-1251-2 (pbk. : alk. paper)
    1. Human Genome Project—Social aspects. 2. Human Genome Project—Moral and ethical aspects.   3. Human genome—Research—Moral and ethical aspects. I. Peters, Ted, 1941– .   II. Series.
QH442.G473   1998
599.93'5—dc21   97-45113
                                                                                    CIP

Contents

PART THREE
Social Challenges

Acknowledgments

THE CHAPTERS OF THIS book were written by a core group of researchers and guest scholars at the Graduate Theological Union in Berkeley, California, as part of a research project sponsored by the Center for Theology and the Natural Sciences and funded by National Institutes of Health Grant #HG00487. The Core Group from 1991 to 1994 consisted of Deborah Blake, R. David Cole, Ronald Cole-Turner, Lindon Eaves, Langdon Gilkey, Philip Hefner, Solomon Katz, Karen Lebacqz, Arthur Peacocke, Ted Peters, Robert John Russell, Thomas A. Shannon, and Roger L. Shinn. The National Advisory Committee and list of consultants included Ian G. Barbour, Paul R. Billings, Hubert Dreyfus, Troy Duster, Henry Greely, Susan Jackson, Mary-Claire King, Margaret McLean, Richard Randolph, Mark Richardson, Robert T. Schimke, and Gunther Stent.

I would like to thank Lisa Dahlen, special project assistant for the CTNS-Human Genome Project, for all her work in helping prepare and edit this manuscript for publication. I am grateful for the diligence, cheerfulness, and attention to detail that she gave to this project. I would also like to thank Fred Sanders for finalizing the manuscript with the publisher.

Contributors

R. David Cole is a molecular biologist and protein specialist, recently retired as professor of molecular and cell biology at the University of California at Berkeley. He has produced 180 scientific papers mostly about chromosomal proteins, and recently, he published the article "Genetic Predestination?" in *Dialog: A Journal of Theology* 33, no. 1 (winter 1994): 17–22.

Troy Duster is professor and former chair of the Department of Sociology as well as director of the Institute for the Study of Social Change at the University of California at Berkeley. He is author of *Backdoor to Eugenics* (Routledge, 1990) and is known for his research into the African American community and problems associated with social stratification.

Henry T. Greely is a professor of law at Stanford Law School in Stanford, California. He is a member of the North American Committee of the Human Genome Diversity Project and chairs its ethics subcommittee.

Philip Hefner is professor of systematic theology and director of the Chicago Center for Religion and Science at the Lutheran School of Theology at Chicago. He is editor of *Zygon* and author of numerous works in the field of theology and natural science including *The Human Factor* (Fortress, 1993).

Karen Lebacqz is the Robert Gordon Sproul Professor of Theological Ethics at the Pacific School of Religion and the Graduate Theological Union in Berkeley, California. The author of two important books on the concept of justice, *Six Theories of Justice* (Augsburg, 1986) and *Justice in an Unjust World* (Augsburg, 1987), she has also edited *Genetics, Ethics, and Parenthood* (Pilgrim Press, 1983).

David A. Peters is professor and chair, Department of Philosophy, University of Wisconsin at River Falls, Wisconsin. He is a specialist in environmental ethics and public health care policy.

Ted Peters, editor of the present volume, is a professor of systematic theology at Pacific Lutheran Theological Seminary and the Graduate Theological Union in Berkeley, California. He served as principal investigator on the project Theological and Ethical Questions Raised by the Human Genome Initiative, sponsored by the Center for Theology and the Natural Sciences and funded by National Institutes of Health Grant #HG00487. He is editor of *Dialog: A Journal of Theology;* and he is author of *GOD—The World's Future: Systematic Theology for a Postmodern Era* (Fortress, 1992), *For the Love of Children: Genetic Technology and the Future of the Family* (Westminster/John Knox, 1996), and *Playing God? Genetic Determinism and Human Freedom* (Routledge, 1997).

Thomas A. Shannon is professor of religion and social ethics in the Department of Humanities, Worcester Polytechnic Institute, Worcester, Massachusetts. He is the editor and author of numerous books in bioethics, such as *Bioethics: Selected Readings* (Paulist, 4th ed., 1993) and *What Are They Saying about Genetic Engineering?* (Paulist, 1985).

Roger L. Shinn is Reinhold Niebuhr Professor Emeritus of Social Ethics, Union Theological Seminary, New York. A distinguished veteran of social ethics, he writes here about the history in which he himself played an important part as the World Council of Churches engaged in research and statement writing on matters of genetic engineering. He is the author of fifteen books including *Forced Options: Social Decisions for the Twenty-first Century* (Pilgrim Press, 3d ed., 1991) and *The New Genetics* (Moyer Bell, 1996).

Laurie Zoloth-Dorfman is associate professor of social ethics and Jewish theology and director of the Jewish Studies Program at San Francisco State University, and is a clinical bioethicist at Children's Hospital in Oakland, California. She is the author of *The Ethics of Encounter: A Jewish Conversation on Healthcare Justice* (University of North Carolina Press, 1998).

[1]

Genes, Theology, and Social Ethics: Are We Playing God?

Ted Peters

NEW KNOWLEDGE GAINED FROM genetics research is raising a host of challenging ethical questions, and these ethical questions are prompting theological reflection. The dramatic scale of the biomedical challenges throws us back upon first principles, back to questions about the nature of human nature, about our relationship to ourselves and to our divine source, God. In the popular press the issue is formulated this way: Are we playing God? A more serious approach seeks to ferret out the implications of the ethical issues and asks, How might theological insights guide and direct ethical deliberation?[1]

The ethical questions emerging from the field of genetics are creating a sense of urgency due to the enormous scale of research associated with the Human Genome Project (HGP), sometimes referred to as the Human Genome Initiative (HGI). Begun in 1988, HGP is a "big science" project, international in scope, involving numerous laboratories and associations of scientists strewn across the landscape, having a current annual U.S. budget of $200 million with a fifteen-year time line and a $3 billion total price tag.[2] The scientific goal is to map and sequence the human DNA. Mapping will eventually tell us the position and spacing of the predicted 100,000 genes in each of the body's cells; and sequencing will determine the order of the four base pairs—the A, T, G, and C nucleotides—that compose the DNA molecule.[3] The primary motive is what drives all basic science, namely, the need to know.[4] The secondary motive is per-

1

haps even more important, namely, to identify the genetic basis for the 4,000 or so genes that are suspected to be responsible for inherited diseases and prepare the way for treatment through genetic therapy. This will be the social benefit, the connection between laboratory and society. The new knowledge will require new thinking about the ethical, legal, and social dimensions of life for the human beings whose cells contain the DNA undergoing studies.

In the case of the HGP, we have scientists who are already aware that their research will have an impact on surrounding society and are willing to share responsibility for it. When James D. Watson counseled the U.S. Department of Health and Human Services to appropriate the funds for what would become HGP, he recommended that 5 percent of the budget be allotted to study the ethical, legal, and social implications of genome research. "We must work to ensure that society learns to use the information only in beneficial ways," says Watson, "and, if necessary, pass laws at both the federal and state levels to prevent invasions of privacy. . . to prevent discrimination on genetic grounds."[5] Watson, who along with Francis Crick is famed for discovering the double helix structure of DNA, was the first to head the Office of Human Genome Research at the National Institutes of Health (NIH). He later resigned amid a dispute with former NIH director Bernadine Healy over the morality of patenting of DNA sequences.[6] Moral controversy has already broken out on this and numerous other issues. Hence, we can expect *genetics* and *ethics* to court each other for the next few years, leading perhaps to a marriage, to *genethics*.[7]

My task here will be to identify and formulate eight of the major ethical issues that are already appearing on the genetic horizon and to bring them into focus by drawing out some of their theological implications.[8] In doing so, I will report on the direction being taken by religious communities in North America that are aware that something significant is happening here. The controversy that broke out over recombinant DNA in the late 1970s and early 1980s precipitated an initial awareness on the part of theologians and other church leaders that research in genetics might require ethical dialogue or monitoring. Since then a handful of theologians and ethicists have begun a serious dialogue with research scientists to sort through the issues. Some ecclesiastical bodies have even prepared official or semi-official statements. These statements provide a valuable resource as we today try to forecast future religious reactions to the developing new knowledge.[9] They also provide a resource for public discussion leading to the eventual formulation of sound social policy.

A statement prepared by the National Council of Churches under the leadership of Roger L. Shinn answers why churches in the United States are concerned about genetic research and therapy, namely, to relieve suffering:

The Christian churches understand themselves as communities dedicated to obeying the will of God through service to others. The churches have a particular concern for those who are hurt or whose faith has been shaken, as demonstrated by the long history of the churches in providing medical care. . . . Moreover, the churches have a mission to prevent suffering as well as to alleviate it.[10]

And with genetic technology in mind, the Church of the Brethren recommends "the continuing use of scientific research for the alleviation of human suffering."[11] What is important here is this: the mission to prevent suffering gives rise to ethical thinking, and ethical thinking is vital to formulating public policy.

1. Genetic Discrimination

One ethical issue on the genetic horizon has already begun to take focus, namely, genetic discrimination.[12] A possible scenario runs like this. In the next few years researchers will identify and locate most, if not all, genes in the human genome that condition or in some cases cause disease. Already we know that the gene predisposing one to cystic fibrosis is found on chromosome 7 and Huntington's chorea on chromosome 4. Alzheimer's disease is probably due to a defective gene on two chromosomes, and colon cancer to one on chromosome 2. The recently discovered predisposition to inherited breast cancer (BRCA 1) is located on chromosome 17 and a second (BRCA 2) on 11, where we also find type one diabetes. Predisposition to muscular dystrophy, sickle-cell anemia, Tay-Sachs disease, certain cancers, and numerous other diseases have locatable genetic origins. More knowledge is yet to come. When it comes, it may be accompanied by an inexpensive method for testing the genome of each individual to see if he or she has genes for such diseases. Testing for many genetic diseases may become routine for newborns just as testing for phenylketonuria (PKU) has been since the 1960s. A person's individual genome would become part of a data bank to which each of us as well as our health care providers would have future access. The advantage is clear: alert medical care from birth to grave could be carefully planned to delay onset, appropriately treat, and perhaps even cure genetically based diseases.

This is an unlikely scenario, however. As good as it may sound, the specter of genetic discrimination may slow it down, if not retard it completely. Broadscale presymptomatic genetic screening is already being positively considered and sharply questioned at the same time. The Institute of Medicine (IOM) argues that mass nonvoluntary screening of newborns is misguided. In the case of

Duchenne muscular dystrophy, for example—an inherited muscle disorder that leads to death during the teen years for which no cure currently exists—such newborn testing can lead only to family grief, asserts the IOM. Our ability to identify genetic diseases will speed ahead while our ability to devise appropriate treatments will lag behind. Families could be beset with knowledge but no cure. The IOM fears that parents of children with known genetic defects might "develop a different outlook on a child destined to die."[13]

More decisively, any grand plan to employ new genetic knowledge for preventive health care is likely to be impeded, if not blocked entirely, for citizens by the U.S. system of financing medical care through commercial insurance. The problem begins with insurability and may end up with a form of discrimination that for genetic reasons prevents certain individuals from obtaining employment and, hence, obtaining medical services. Even with a government-regulated program of access to basic health services, the need to purchase supplemental insurance to cover serious diseases makes many of us with certain genetic configurations vulnerable to discrimination.

Insurance works by sharing risk. When risk is uncertain to all, then all can be asked to contribute equally to the insurance pool. Premiums can be equalized. Knowing the genetic disorders of individuals, however, could justify higher premiums for those demonstrating greater risk. The greater the risk, the higher the premium. Insurance may even be denied those whose genes predict extended or expensive medical treatment.[14]

For three-quarters of U.S. citizens, medical insurance is tied to employment. Among *Fortune*'s 500 top companies, twelve already report using genetic testing for employment purposes. Although testing in the past was justified initially for public health purposes, increasingly employers may be motivated to use testing to cut premium costs for the medical insurance they pay on behalf of employees. Underwriters already deny or limit coverage to gene-related conditions such as sickle-cell anemia, atherosclerosis, Huntington's disease, Down syndrome, and muscular dystrophy.[15] This list could increase. Individuals with genetic predispositions to expensive diseases may become unemployable, uninsured, and finally unable to acquire medical care. Between thirty and forty million persons in the United States currently live with insufficient or no medical coverage; despite its promises for a better life, HGP could inadvertently add a whole new class of poor people.

This scenario may happen gradually, unevenly, and unnecessarily. Geneticists estimate that each of us carries five to seven lethal recessive genes as well as a larger number of genes that make us susceptible to developing multifactorial diseases. There is probably no one whose genome is disease free. Yet this may not be obvious in the initial years that society wrestles with the problem. Those who are tested in the early years will suffer discrimination because of their apparent singularity. Although it will never be the case that all people will face identical genetic risk, eventually, we are likely to find that differences are minimal. Only later, when we discover again the relative equality of risk distribution, will the pressure for stigma be released.[16] However, this may come only after considerable social damage has been done.

In the meantime, ethicists operating out of rights' theory are seeking protection from discrimination by invoking the principles of confidentiality and privacy. They argue that genetic testing should be voluntary and that the information contained in one's genome be controlled by the patient. This argument presumes that if information can be controlled, then the rights of the individual for employment, insurance, and medical care can be protected. There are some grounds for thinking this approach will succeed. Title VII of the 1964 Civil Rights Act restricts preemployment questioning to work-related health conditions, and its paragraph 102.b.4 potentially protects coverage for the employee's spouse and children. Current legislative proposals seem to favor privacy.[17]

Nevertheless, I believe the privacy defense can at best be a mere stopgap effort. In the long run, it will fail. Insurance carriers will press for legislation more fair to them, and eventually, protection of privacy may slip. In addition, the existing state of computer linkage makes it difficult to prevent the movement of data from hospital to insurance carrier and to anyone else bent on finding out. In my judgment, the most important factor is the principle that genome information should not finally be restricted. The more we know and the more who know, the better the health care planning can be. In the long run, what we want is information without discrimination. The only way to obtain this is to restructure the employment–insurance–health care relationship. The current structure seems to make it profitable for employers and insurance carriers to discriminate against individuals with certain genetic configurations— that is, it is in their best financial interest to limit or even deny health care. A restructuring seems called for so that it becomes profitable to deliver, not withhold, health care. To accomplish this, the whole nation will have to become more egalitarian—that is, to think of the nation itself as a single community willing to care for its own constituents.[18]

Where is the nation's church leadership now that this debate is beginning to heat up? It is clear that religious ethicists oppose genetic discrimination. In 1989, the Church and Society Commission of the World Council of Churches released a study document, *Biotechnology: Its Challenges to the Churches and the World*, which draws attention to "unfair discrimination . . . in work, health care, insurance and education."[19] Similarly, in the proposal approved by its 1992 General Conference, the United Methodist Church Genetic Task Force listed prominently among possible HGP repercussions, "discrimination: the suffering and/or hardship that may result for persons with late-onset disease like Huntington's or Alzheimer's disease, or with a genetic predisposition to diseases like high cholesterol levels or arteriosclerosis."[20] And in 1989, the Seventeenth General Synod of the United Church of Christ meeting in Fort Worth, Texas, approved a pronouncement that included a rejection of "screening as a basis for determining civil, economic, or reproductive rights."[21] A resolution passed at the Seventieth General Convention of the Episcopal Church in July of 1991 states forcefully: "The use of results of genetic screening of adults, newborns and the unborn for the purpose of discrimination in employment and insurance is unacceptable." This clear stand against genetic discrimination provides a solid foundation from which to build an ethical proposal; but it stops here. There are hints that church ethicists will side with those who advocate privacy; and there are hints that they favor some sort of national program guaranteeing health care to everyone. What we do not see as yet among religious leaders is any overall vision regarding the potential value (or nonvalue) of widespread use of genome information for health care delivery.

2. The Abortion Controversy Intensifies

Perhaps the most divisive moral issue in the United States is the practice of abortion. The advance of genetic knowledge and the development of more sophisticated reproductive technologies will only add nuance and subtlety to an already complicated debate. Techniques have been developed to examine in vitro fertilized (IVF) eggs as early as the fourth cell division in order to identify so-called defective genes such as the chromosomal structure of cystic fibrosis. Prospective parents may soon be able to fertilize a dozen or so eggs in the laboratory, screen for the preferred genetic makeup, implant the desired zygote(s), and discard the rest. What will be the status of the discarded preembryos? Might such disposals be considered abortions? By what criteria do we define "defective" when considering the future of a human being? Should prospective parents limit themselves to eliminating children with defects, or should they go on to test for positive genetic traits such as blue eyes and higher intelligence? If so, might this lead to a new form of eugenics, to selective breeding based upon personal preference and prevailing social values?[22] What will become of human dignity in all this?

The challenges compound quickly. For the sake of simplicity at this point, let us limit our second ethical issue to that of selective abortion. University of Texas law professor John A. Robertson says that the central ethical issue in preimplantation genetic testing is the acceptability of discarding unwanted embryos. Those who view the fertilized egg as the point at which human dignity and human rights begin have argued that all such embryos must be placed in the uterus and given the opportunity to grow. Those holding this view would be consistent in opposing the discarding of embryos with genetic defects just as they oppose abortion of fetuses with the same defects. They might in turn support legislation that would prohibit embryo discard. In contrast, those favoring preimplantation testing with its accompanying discard of unwanted zygotes would argue that the embryos are too rudimentary in form or development to be owed respect, since they are still undifferentiated cells that are not yet clearly individual.[23]

It may be relevant to mention that the legal precedent set by *Roe v. Wade* would not legitimate discarding preimplanted embryos. This Supreme Court case legalized the use of abortion to eliminate a fetus from a woman's body as an extension of a woman's right to determine what happens to her body. This would not apply to preimplanted embryos, however, because they are life-forms outside the woman's body.

Still open for debate is the question: Does the discarding of nonimplanted embryos constitute abortion? Perhaps a more pressing and related concern will soon be selective abortion. As more and more genes predisposing to disease are located, more and more genetic tests will be developed. These tests may become applied routinely to pregnancies. Knowing that eight out of ten positive prenatal tests for Down syndrome and cystic fibrosis result in decisions to abort, we can forecast an increase in abortion decisions on genetic grounds. We can imagine prenatal genetic tests numbering in the hundreds. Every pregnancy will become virtually tentative until such tests are run.

The ethical question—by what criteria do we deem a genetically defective or undesirable fetus abortable?—similarly was not addressed by *Roe v. Wade.* The present practice of abortion by choice prior to the third trimester locates the choice in the pregnant woman, but it does not provide ethical criteria for distinguishing better from worse choices.

The Roman Catholic tradition has set strong precedents regarding the practice of abortion; and the Vatican is beginning to apply them to genetics and related reproductive issues. The precedent against aborting the unborn is clear from the Second Vatican Council: "From the moment of its conception life must be guarded with the greatest care, while abortion and infanticide are unspeakable crimes."[24] The challenge to ethicists in the Roman Catholic tradition in the near future will be to examine what transpires at the preimplantation stage of the embryo (or preembryo) to determine if the word "abortion" applies. The thinking of the Congregation for the Doctrine of the Faith seems to lead in this direction: "The human being is to be respected and treated as a person from the moment of conception," therefore "the embryos . . . must be respected just like any other human person."[25] Pope John Paul II says the fertilization and discard of nonimplanted embryos are morally unacceptable.[26]

This line of thinking may lead to recommending that genetic testing be pushed back one step farther, to the gamete stage prior to fertilization. The genetic makeup of sperm and ovum separately could be tested, using acceptable gametes and discarding the unacceptable ones. The Catholic Health Association of the United States pushes back yet another step by recommending the development of techniques of gonadal cell therapy to make genetic corrections in the reproductive tissues of prospective parents long before conception takes place—that is, gametocyte therapy.[27] Other issues such as the criteria for genetic acceptability remain, to be sure, but the problem of postconception abortion may be solved in this way.

Protestants have comparatively less to say about the implications of abortion ethics to genetics or vice versa. For example, the 1991 "Social Statement on Abortion" by the Evangelical Lutheran Church in America rehearses many of the issues already argued in the 1970s without taking a stand regarding the morality of abortion by choice.[28] The Methodist task force recognizes the value of genetic testing in prenatal diagnosis, but no connection is drawn to the abortion controversy. The World Council of Churches draws the connection and asks, "When does individual life begin?" Not at fertilization. At this early stage, the WCC document says, "the specific interrelationships among cells that later characterize the whole organism . . . are not yet permanently established."[29] Such an understanding could logically lead to approval of the discard of genetically unacceptable embryos, but that is not directly asserted. The United Church of Canada grants moral approval to abortion when it is the "lesser of two evils," but it offers no precise application to genetic testing.[30] The United Church of Christ statement supports "genetic screening of pregnancies at risk," but that is as close as it gets to making the connection between abortion and genetics.[31] Ronald Cole-Turner's commentary on the UCC statement applies to Protestants in general: "When it comes to the central moral issue—whether or not to terminate a pregnancy if a defect is identified—these professionals tend to be non-directive. . . . Does the prediction of a genetic disease call for an abortion?"[32] As yet there is no answer.

3. Patenting God's Creation

One might think that the controversy over whether we should patent knowledge of DNA sequences would be limited to an argument among scientists, biotechnology companies, and the government. Yet religious voices have been raised. The louder religious voices have shouted "No!" to patenting intellectual property regarding genes on the grounds that DNA belongs to God's creation. Where are we on this? Let me try to explain briefly this third issue.

The 1990s chapter in the larger story of the controversy over patenting life-forms began with the initial filing in June 1991 for patent property rights on 337 gene fragments, and a second filing in February 1992 on 2,375 more partial gene sequences by J. Craig Venter. Venter at that time was pursuing research at the NIH's National Institute of Neurological Disorders and Strokes. His aim was to locate and sequence the 30,000 or so complementary DNAs, or cDNAs, from the human brain.

Venter's method, based on a deceptively simple insight, was key. The task of the HGP has been to sequence the entire three billion nucleotides in the DNA and to locate where on the DNA the genes are sited. Relatively speaking, only a small portion of the DNA functions as genes, about 3 percent. The nongenetic material has been affectionately labeled "junk DNA." If one wants to find only the genes, thought Venter, then why bother with plodding through all the junk DNA? Noting that only the genes, not the junk DNA, code for proteins by creating messenger RNA (mRNA), Venter set his focus on mRNA. He began making sturdier clone copies of the otherwise fragile mRNA; and the stronger and analyzable copies he called cDNAs. By sequencing only the cDNAs, he could be assured that he was gaining knowledge of actual genes; and by focusing the research that way, he brought the price of sequencing down dramatically.

These clones made from mRNAs represent a part of the coding region of genes on the DNA chain. By sequencing a short stretch of cDNA clones—about three hundred to five hundred bases and not necessarily the entire gene—Venter created what he called an "expressed sequence tag," or EST. Venter had begun using automatic sequencing machines to the limit of their capacity and was churning out 50 to 150 such tags per day.

In the fall of 1992 the U.S. Patent and Trademark Office made a preliminary ruling that denied the applications on the grounds that gene fragments could not be patented without knowing the function of the gene. Then NIH director Bernadine Healy pressed for an appeal. When Harold Varmus took over the helm from Healy, he withdrew the applications, saying that patents on partial gene sequences are "not in the best interests of the public or science."[33] Thus, the ball entered the court of the private sector where similar patent applications were filed. Some were filed by Venter and his colleagues after he left NIH, having garnered $70 million in venture capital to start a private biotech company, The Institute for Genomic Research (TIGR).[34] The result has been controversy with numerous hotly debated questions.

Two of the many hotly debated questions are these: Should patents be granted for knowledge of gene sequences at all? And, if so, should they be granted to a government-funded agency such as NIH or only to the private sector? The first question regarding the patentability of genetic knowledge can be broken down into three subquestions. We will not take up the question whether the government or the private sector should own the patents because the owner of the patent is irrelevant to what has become the religious issue.

The first subissue is this: Does successful cDNA sequencing count as patentable knowledge about the genes themselves? Venter admitted that even though he could tag a cDNA sequence, he had no idea what it does, unless it belongs to a sequence from a gene whose function is already known. James D. Watson, then head of the National Center for Human Genome Research at NIH, vitriolically opposed this rush to patenting, decrying the overvaluing of what was being accomplished.[35] Sequencing a short piece of an unidentified clone with an automated sequencing machine is a "dumb, repetitive task."[36] What remains to be done, and what is decisive, Watson said, is to interpret the data so that we learn exactly what function each gene performs. Similarly, the Gene Patent Working Group, an interagency committee set up by the White House Office of Science and Technology Policy (OSTP), declared that ESTs are merely research tools and should not be granted patent protection that belongs to knowledge of the complete gene sequence with its function.[37] In sum, critics argued that patenting at this early stage was premature.

The second subissue is this: Should this particular intellectual knowledge be patentable? To patent, an invention must meet three criteria: it must be novel, it must be nonobvious, and it must have utility. Certainly, Venter's application was novel because the newly identified genes seem not to have been sequenced before. However, it was less obvious that it was nonobvious. Venter made use of existing technology; he did not invent it. Nevertheless, Venter contributed an insight that led to unprecedented speed in scientific searching ability. That must be worth something, his sponsoring lawyers argued. Even so, the toughest hurdle, utility, remained. Venter's genes were dubbed "naked," meaning that their function was unknown. Until the function could be known, no movement could be made toward developing medical or other benefits.[38]

The third subissue brings us to a more philosophical question: Should intellectual knowledge regarding natural processes in principle be patentable? Does witnessing an existing natural phenomenon in itself warrant patent protection for the witness? Should an astronomer be able to patent every new galaxy he or she discovers? No, would be the answer someone like Justice William O. Douglas would give. Writing for the majority in the 1948 U.S. Supreme Court case of *Funk Brothers Seed Co. v. Kalo Inoculant Co.,* he wrote, "Patents cannot issue for the discovery of the phenomena of nature. . . . [Such] are manifestations of laws of nature, free to all men and reserved exclusively to none."[39]

Are cDNAs a natural phenomenon or a human invention? The cDNA is not a gene per se. Rather, it is a copy version of a gene with the introns edited out. It does not occur naturally. It is coded into messenger RNA by the process that reads the raw cellular DNA. This fact leads to an interesting double-sidedness on the part of Daniel Kevles and Leroy Hood. On the one hand, they argue: "Since it can be physically realized by a devising of human beings, using the enzyme reverse transcriptase, it is patentable." On the other hand, cDNA tells us what is in the original genomic DNA. So, Kevles and Hood are troubled: "If anything is literally a common birthright of human beings, it is the human genome. It would thus seem that if anything should be avoided in the genomic political economy, it is a war of patents and commerce over the operational elements of that birthright."[40]

In 1995 the Human Genome Organization (HUGO) issued a statement opposing the patenting of cDNAs because it would impede the free flow of scientific information:

> HUGO is worried that the patenting of partial and uncharacterized cDNA sequences will reward those who make routine discoveries but penalize those who determine biological function or application. Such an outcome would impede the development of diagnostics and therapeutics, which is clearly not in the public interest. HUGO is also dedicated to the early release of genome information, thus accelerating widespread investigation of functional aspects of genes.[41]

The Craig Venter affair draws a more hysterical interpretation by Andrew Kimbrell, policy director of the Washington, D.C.-based Foundation on Economic Trends.

> The entire human genome, the tens of thousands of genes that are our most intimate common heritage, would be owned by a handful of companies and governments. If Venter's applications or those of others are accepted, in a short time a few government bureaucracies and powerful corporations will have a monopoly on the use and sale of all human genes.[42]

In sum, cDNAs may prove patentable on the grounds that they are the product of a humanly devised process of gaining intellectual knowledge. However, as long as the Douglas principle holds that processes already occurring in nature are exempt, the human genome itself will not become patentable.

This became a religious issue on May 18, 1995, at a Washington, D.C., press conference where it was announced that religious leaders representing more than eighty groups had signed a statement opposing patenting. Numerous Roman Catholic bishops along with Jewish, Protestant, Islamic, Hindu, and Buddhist leaders signed the following statement:

> We, the undersigned religious leaders, oppose the patenting of human and animal life forms. We are disturbed by the U.S. Patent Office's recent decision to patent human body parts and several genetically engineered animals. We believe

that humans and animals are creations of God, not humans, and as such should not be patented as human inventions.[43]

The press conference, called the Joint Appeal Against Human and Animal Patenting, included Rabbi David Saperstein, director of the Religious Action Center of Reform Judaism; Abdurahman Alamoudi, executive director of the American Muslim Council; Wesley Granberg-Michaelson, secretary general of the Reformed Church in America; Richard Land, executive director of the Christian Life Commission of the Southern Baptist Convention; and Kenneth Carder and Jaydee Hanson of the United Methodist Church. "By turning life into patented inventions," said Jeremy Rifkin, whose Foundation on Economic Trends orchestrated the event, "the government drains life of its intrinsic nature and sacred value."

"Marketing human life is a form of genetic slavery," Richard Land was quoted in papers across the country. "Instead of whole persons being marched in shackles to the market block, human cellines and gene sequences are labeled, patented and sold to the highest bidders." Land added a prophetlike judgment against playing God in the laboratory: "We see altering life forms, creating new life forms, as a revolt against the sovereignty of God and an attempt to be God."[44]

This event marks a point of public meeting between the religious and scientific communities, a meeting that quite unfortunately has the appearance of a battle. That the religious community would enter into a battle over patenting is most unfortunate, in my judgment. The main problem with the statement is that it is vague and inflammatory. The statement fails to make clear what the Patent and Trademark Office strives to make clear, namely, it does not grant patents to natural phenomena, such as human beings or animals or body parts in any ordinary sense. It grants patents to human inventions that are novel, nonobvious, and useful.

The theology presupposed here is at odds with religious traditions such as Christianity and Judaism that believe in a transcendent and holy God who created the natural world, loves it, and asks that the human race created in the divine image strive to make the world a better place. These traditions hold that the creator God is sacred, not the creation. The May 18 theology reflects, rather, the point of view of Rifkin, famed for his outspoken resistance to progress in biological research and medical technology. This is a tacit naturalism of sorts. In his book *Algeny*, he describes his mission as a "resacralization of nature."[45]

Sweeping over the difference between these respectively different metaphysical positions could have ethical ramifications. Jews and Christians hold that we humans should be good stewards of our God-given creativity and out of love for neighbor pursue, among other things, the development of better ways to relieve human suffering and improve the human experience with life. The Rifkin position implies that nature prior to human creative intervention is sacred and should be left alone. This severs the warrant for pursuing medical research and development of therapies that could relieve human suffering and improve the health of the human race. Here is the issue on which I believe the patent debate should focus: How can patents help or retard the development of genetically based therapy for cancer, heart disease, cystic fibrosis, Alzheimer's, Huntington's disease, Williams syndrome, and countless others? In effect, the religious leaders have unnecessarily cut themselves off from making a contribution to this central concern. We will take up this concern later when we address the question: Should we play God?

4. Cloning

On February 23, 1997, the world woke up to the fact that the era of cloning had dawned. At the Roslin Institute near Edinburgh, Scotland, embryologist Ian Wilmut produced a live adult lamb from cells originating in a sheep mammary gland. The method was simple, technologically speaking. Wilmut took a mammary cell from an adult sheep and placed its DNA into the egg of another sheep. He removed the egg's DNA and fused the adult DNA to the egg. The fused cell began to grow and divide, just like a normally fertilized egg. It became an embryo, it was planted in the womb of a ewe, and at the time of publication in February 1997 it was already a seven-month-old lamb named Dolly. DNA tests show that Dolly contains only the genes of the adult ewe who provided her DNA.

What are the implications? We must distinguish between animal cloning and human cloning. We can fully expect commercial interest to expand into cloning technology. The most milk-producing cows and best-tasting meat-bearing steers may find themselves cloned by the thousands. Cloning may replace breeding for reproducing the fastest racehorses. Medical researchers such as Wilmut may rely in part upon cloning for health care products.

Some theologians worry about what such a technology might do to our attitude toward animals. The replication of identical animals by the thousands seems intuitively to be an assault on the respect we hold for other sentient beings. Spokespersons for the Society, Religion, and Technology Project of the Church of Scotland immediately went on the Internet to plead for caution. The Bible pictures a wonderfully diverse creation whose sheer diversity is a cause of praise to the Creator, the Church of Scotland said. With this assumption, to produce replica animals on demand would go against something basic and God-given about nature and life.

In contrast, I do not find the replication of animal genomes carrying this much theological freight. Although we can provide scientific justification for the value of diversity in nature, I see no biblical or theological grounds for celebrating diversity per se. Diversity may strengthen the gene pool, according to evolutionary theory; but it is an unwarranted jump to assert that this is God's will or that nature should be praised for it. Taken to its extreme, this argument should have been raised over the centuries against farmers engaged in selective breeding. But, to my knowledge, selective breeding that reduces diversity while creating new genomes has not even stimulated interest among theologians, let alone proscription. The anticloning appeal to diversity seems to be a newly invented criterion invoked for just this case.

The issue for the future will be the question: Should we clone human beings? The initial public and ecclesiastical reaction has been a resounding "No!" Prior to the birth of Dolly, four nations had already outlawed human cloning: Great Britain, Spain, Denmark, and Germany. Many voices in the United States immediately demanded a similar ban. President William Clinton referred the matter to the National Bioethics Advisory Committee. In Scotland, Donald Bruce, who directs the Society, Religion, and Technology Project, described human cloning as a "perversity." To use technology to replicate a human being is against the basic dignity of our uniqueness in God's sight, Bruce argued; it would be ethically unacceptable as a matter of principle because it violates the uniqueness of the life that God has given to each of us and to no one else.

My considered judgment is that, in principle, no distinctively theological affirmations would make cloning humans unethical. Though not unethical, it might be unwise. I favor a temporary moratorium against human cloning until good reasons for it can be brought forth and considered. I do not favor an absolute ban that eliminates all future consideration.

The argument raised by the Church of Scotland fits with fears enunciated by many, namely, that cloning would compromise human identity and violate human dignity. This is an unfounded fear, in my judgment. It is based on the gene myth—which we will discuss later—according to which "it's all in the genes." Widespread is the assumption that who we are is determined by our genetic code, that DNA determines our destiny. With this assumption we can see why some might feel their identity would be compromised when another person shares the same genome.

Yet neither science nor common sense supports this assumption. Scientifically, the genes alone do not determine our identity. Who we are is influenced by our DNA, to be sure; but how the genes behave is influenced by environmental factors in growing up. In addition, common sense gained from everyday observation reveals that no matter how much two people share in common, they still differ. Who we are as individual persons is determined by three things:

our genome, our environmental influences, and the appearance of a subjective self with free will and the ability to engage in self-definition. Even if we granted that God loves diversity and also granted that individual identity is important, cloning would be at most only a partial threat.

The experience of identical twins or triplets is informative. To be identical, siblings must have the same genome. Yet each twin grows up with his or her own subjectivity and sense of identity, and he or she can claim his or her individual rights. The experience of a cloned person would be similar. He or she would be aware that another person shares the same genetic code and might even find this fascinating; nevertheless, he or she would be just as much an individual as any of the rest of us.

It would be society's moral obligation to treat cloned persons as individuals as well. I hold theologically that *God loves each of us regardless of our genetic makeup, and we should do likewise.* In secular language, each of us should be treated with dignity. I would not like to see the fear that cloned persons have less identity become translated by society into a stigma in which such persons are denied dignity.

5. Genetic Determinism, Human Freedom, and the Gene Myth

The fifth issue deals with the problem of myth and reality, of public perception and laboratory truth. It has to do with the growing popular image of the almighty gene as the all-determining factor in the human condition. It begins innocently with the thought that if we could only find the gene for a certain disease, then we could find the cure by simply manipulating this gene. We then continue with the logic. Why stop with diseases? Do genes also determine behavior? If so, should we blame persons for their antisocial behavior or judge them as innocent? Should we try to alter the genes of antisocial individuals or groups? "It's all in the genes," we say.

This line of thinking belongs to what I call the gene myth, namely, a widespread cultural thought form that says, "It's all in the genes." The gene myth is deterministic in two senses. The first sense is puppet determinism, wherein we assume the DNA acts like a puppeteer and we dance on genetic strings like a puppet. If the DNA determines our hair color and what diseases we will have, then perhaps the DNA determines how we will behave and may even control our virtues and vices. The second is Promethean determinism, wherein we assume that once our scientists have learned how DNA works, we can take charge—that is, we can get into the DNA with our wrenches and screwdrivers and modify it so as to guide our evolutionary future. Puppet determinism presumes that we are victims of our genes, whereas Promethean determinism presumes that we can take charge of our genes. Both belong to the gene myth of our time. Both may be misleading.

Why might the gene myth be misleading? Because, for the most part, molecular biologists see little or no evidence supporting a philosophy of genetic determinism that would alter our conventional understanding of human freedom. At minimum, nurture remains as important as nature. In opposing what he calls "genetic predestination," in a *Nature* editorial, John Maddox draws us back to the basics: genotype underdetermines phenotype.

> The link between genotype and phenotype is not always unambiguous. A genotype may be a necessary but not a sufficient, condition for the phenotype; the individual concerned inherits only a susceptibility for the phenotype . . . it would be rash to deny that the missing ingredients may be aspects of nurture . . . . Is it likely that some of the most labile aspects of the human phenotype, those constituting personality, will be exempt from similar and more subtle influences?[46]

Such deliberation leads molecular biologist R. David Cole to write against genetic predestination:

> There is no reason for the non-scientist to be intimidated by the success of the deterministic approach in elucidating the biological role of genes in human nature, and certainly no reason to be intimidated by any scientist who might try to convince us that determinism is all that is. Although the case for free will cannot be rigorously proven, those of us who believe in it need feel no threat from the findings of the Human Genome Initiative.[47]

Yet there is reason to worry on another but related front: the misconception of gene power may have deleterious social consequences. This misconception has been labeled by some the "Strong Genetic Principle," by some "genetic essentialism," and by others the "gene myth."

The *gene myth* emerges from two sources: the scientists themselves and the press reporting on the science. Some scientists are tempted to believe the gene myth because of a failure to distinguish between methodological reductionism and ontological reductionism. Reductionism as a method that focuses on the genes as the object of research has been extremely valuable in the production of new knowledge. No one would want to surrender it. Yet to extrapolate on the success of what is being learned about genes to a philosophical worldview—a worldview that says if we only keep studying DNA, we will find that "it's all in the genes"—commits the fallacy of hasty generalization. Human being—that is, human ontology—may very well be the result of numerous factors, only one of which is DNA. To know for sure, we need to study humanity at all levels, from the cell on up through physiology to society and even spirit. Ontological reductionism also misleads interpreters of the science, such as the press. The press looks for headlines emphasizing breakthroughs and revolutions just around the corner, and this attracts headlines that tell us "it's all in the genes!"

In the early days of the HGP, rightfully hopeful spokespersons left the impression that genetic knowledge would be all-explanatory. Walter Gilbert said that "to identify a relevant region of DNA, a gene, and then to clone and sequence it is now the underpinning of all biological science."[48] James Watson told *Time* magazine, "We used to think our fate was in our stars. Now we know, in large measure, our fate is in our genes."[49] The DNA sequence has been called the "Rosetta Stone" and the "Holy Grail" and even "the ultimate explanation of human being." All of this upsets Harvard Medical School researcher Jonathan Beckwith. This sort of talk raises "a false hope of cures for disease" that may not come. In addition, "genetic explanations for intelligence, sex role differences, or aggression lead to an absolving of society of any responsibility for its inequities, thus providing support for those who have an interest in maintaining those inequities."[50] The danger with the "it's all in the genes" philosophy is that it might provide an excuse—an illegitimate excuse—for withdrawing proper attention to social responsibility. In their diatribe *Exploding the Gene Myth*, Ruth Hubbard and coauthor Elijah Wald make the point bluntly:

> Many new genetic breakthroughs . . . do not make people healthier; they merely blame genes for conditions that have traditionally been thought to have societal, environmental, or psychological causes. . . . The myth of the all-powerful gene is based on flawed science that discounts the environmental context in which we and our genes exist. . . . What I object to is the reductionist effort to explain living organisms in terms of the workings of important molecules and their component parts.[51]

The gene myth draws our attention to an ominous cloud on the social horizon, namely, the possible linkage of several things: genes, crime, class, and race. Let us look first at the relationship between genetic determinism and the question of criminal guilt or innocence, and then turn to the possible repercussions for racial justice.

Does a genetic predisposition to antisocial behavior make a person guilty or innocent before the law? Over the next decade our legal system will have to contend with a rethinking of the philosophical planks on which we have constructed concepts such as free will, guilt, innocence, and mitigating factors. There is no question that research into the connection between genetic determinism and human behavior will continue and new discoveries will become immediately relevant to the prosecution and defense of persons accused of crimes.

Threatened here is our concept of free will. The assumption of the Western philosophy coming down to us from Augustine that underlies our understanding of law is that guilt can be assigned only to a human agent acting freely. Will the concept of genetic predisposition compromise our confidence in free agency? The specter on the genetic horizon is that confirmable genetic predispositions to certain forms of behavior will constitute compulsion, and that will place us at a fork in the legal road: either we declare the person with a genetic predisposition to crime innocent and set him or her free, or we declare him or her so constitutionally impaired as to justify incarceration and isolation from the rest of society. The first fork would jeopardize the welfare of society; the second fork would violate individual rights.

The courts have already encountered defense claims of innocence on the grounds that violent behavior was due to an XYY chromosome genotype or due to genetic predisposition to alcoholism.[52] The genetic defense has sought to "excuse" an accused person from responsibility on the grounds that he or she is compelled by what is in the DNA to engage in criminal activity. To date the courts have by and large found this argument unconvincing. The genetic defense is not strong enough to exculpate a person from guilt; but the courts have on occasion used genetic reasoning for mitigating the sentence. In *Baker v. State Bar of California* in 1989, for example, the defendant was found guilty of embezzlement despite his claim that he was compelled to steal because of his genetic predisposition to alcoholism. Yet the alcoholism defense provided sufficient reason to lighten his sentence.[53]

The courts usually find the genetic excuse unconvincing for two reasons. First, the state of scientific research is as yet unable to prove a direct causal link between an individual genome and specific behaviors. Correlation studies are insufficient to constitute proof of guilt in a legal setting. The operative assumption here is that genetic predisposition is only one factor among others in determining behavior—that is, both nature and nurture are determining factors. If in the future science can demonstrate a direct genetic cause for a specific criminal act, then the matter will deserve further consideration. Second, the courts distinguish sharply between a genetic predisposition and our knowledge of such a genetic predisposition, assuming that such knowledge provides a person with the free will necessary to control behavior. In the case of *Baker v. California* just mentioned, the court focused on the relation of genome to free will. The true mitigating factor was not Baker's genetic predisposition; it was the fact that Baker did not know at the time of the embezzlement that he had this genetic condition. Once Baker learned about his biological propensity to alcoholism and in turn to steal, the court declared that from now on he could be held responsible for his behavior. It presumed Baker could and should take appropriate steps to overcome his precipitating condition and to prevent misconduct in the future.

If we remind ourselves of what John Maddox and R. David Cole said regarding the relation between the genotype and the phenotype and the present lack of threat from genetic predestination, then it would be premature to think in terms of excusing criminal behavior on the basis of DNA. After examining the existing court precedents, Maureen P. Coffey concludes that

> recognition of biological determinism need not require the adoption of a constitutional or special genetic defense. Because an individual may be *more vulnerable* to developing a chemical addiction or an antisocial personality disorder does not mean that individual *in fact* will develop those conditions, or that the individual has absolutely *no control* over such development. The existence of a genetic condition merely provides more insight into whether a person possessed enough free will or rational ability to control and understand her behavior.[54]

If eventually we could show scientifically that "it's all in the genes," however, then Coffey would argue for incarceration on the grounds that society needs to be protected from criminal behavior: "While incarceration of a genetically afflicted offender would not serve the traditional objectives of retribution and deterrence, society nonetheless may determine that the isolation function outweighs these."[55] That is, the court could decide to restrict the social freedom of an individual if his or her genome identifies a threat to society.

That society needs to be protected from criminal behavior, and that such protection could be had by isolating individuals with certain genetic predispositions, leads to further questions regarding insanity and race. The issue of insanity arises because we can predict that the genetic defense may rely upon precedents set by the insanity defense for diminished capacity. The courts treat insanity with a focus on the insane person's inability to distinguish right from wrong when committing a crime. When a defendant is judged innocent on these grounds, he or she is incarcerated in a mental hospital until the medical evaluators judge that the individual is cured. Once cured, then released. In principle, such a person might never be judged "cured" and spend more time in isolation than the prison penalty prescribed for the crime, maybe even the rest of his or her life. Should the genetic defense tie itself to the insanity defense, and if one's DNA is thought to last a lifetime, then the trip to the hospital may become the equivalent of a life sentence. In this way the genetic defense may backfire. When it backfires, it may be picked up by the prosecution for the purpose enunciated by Coffey, namely, to incarcerate certain individuals with certain genomes.

What if this type of thinking becomes applied to groups rather than merely individuals? What if genes apparently shared by a race become the criteria of something akin to incarceration, namely, discrimination? The controversy over the book *The Bell Curve* may provide a bellwether. Authors Charles Murray and Richard Herrnstein employ not molecular biology but statistics gained from IQ scores to develop arguments that extend the gene myth. They argue that intelligence plays a decisive role in civil life in the United States. Low cognitive ability contributes to social problems such as poverty, school dropout rates, illegitimate births, unemployment, welfare, and crime. They then rank relative cognitive ability by groups—ethnic and racial groups—identifying higher IQs among Jews, Asians, and whites while identifying lower IQs among Latinos and African Americans. When public policy implications are drawn from such scientific data, the recommended social program would shuttle greater financial resources toward the cognitive elites and remove support for the cognitively challenged on welfare or in Head Start programs.[56]

The gene myth alarms sociologist Troy Duster. Duster worries about racial repercussions if we identify class or crime with genes and then genes with race because we may inadvertently provide biological support for prejudice and discrimination. He takes cognizance of the disproportionately large number of African American men in U.S. prisons. This is a social issue, not a biological one, he insists. He sounds the alarm: "Today, the United States is heading down a road of parallel false precision in this faith in the connection between genes and social outcomes. This is being played out on a stage with converging preoccupations and tangled webs that interlace crime, race, and genetic explanations."[57]

For the most part, church leaders and theologians have not kept abreast of this issue within the wider debate over genetics; at this point we await theological discussion. Such theological discussion should deal with two matters, social justice and human sin. Elsewhere in this volume Karen Lebacqz will take up the social justice dimension. Here we turn to the matter of genes and sin.

6. The Gay Gene and the Question of Original Sin

Paralleling what we said in the context of law, we now ask, Does a genetic predisposition to what many believe to be immoral behavior make us innocent or guilty? As just noted, the scientific community and the legal community by and large reject the idea of any deterministic link between genes and behavior for two reasons: either there are insufficient data to prove such a link or, in principle, free human beings are said to transcend their genes. Although the scientific jury is still out awaiting further data, we can certainly forecast that new knowledge may bring new considerations. As an example of what we might eventually face, here we will track the trajectories of ethical and theological considerations emerging from the 1993 discovery of a possible genetic disposition to male homosexuality.

The question here is this: Given that some religious traditions have hitherto thought of homosexual activity as sinful, might scientific evidence that such a sexual predisposition is found in the genes render it nonsinful? Can new scientific knowledge change theology and ethics? It is not my task here to take a stand on the morality or immorality of homosexual relations. Rather, I wish to track the possible impact of new discoveries on moral thinking. Fascinating here is the question: Does a genetic predisposition toward behavior deemed unacceptable look at all like original sin?

In the summer of 1993 a scientific bomb exploded that is still causing considerable ethical fallout. Dean H. Hamer and his research team at the National Cancer Institute announced that they discovered evidence that male homosexuality—at least some male homosexuality—is genetic. Constructing family trees in instances where two or more brothers are gay combined with actual laboratory testing of homosexual DNA, Hamer located a region near the end of the long arm of the X chromosome that likely contains a gene influencing sexual orientation. Because men receive an X chromosome from their mother and a Y from their father (women receive two X's, one from each parent), the possible gay gene is inherited maternally. Mothers can pass on the gay gene without themselves or their daughters being homosexual. A parallel study of lesbian genetics is as yet incomplete; and the present study of gay men will certainly require replication and confirmation. Scientists do not yet have indisputable proof. We still await replication and confirmation. Nevertheless, Hamer was ready to write in the article making the dramatic announcement, "We have now produced evidence that one form of male homosexuality is preferentially transmitted through the maternal side and is genetically linked to chromosome region Xq28."[58]

The press erupted with follow-up articles, the kind of articles that spread the gene myth. *Time* magazine projected an ethical and political forecast: "If homosexuals are deemed to have a foreordained nature, many of the arguments now used to block equal rights would lose force." *Time* cited a gay attorney saying, "I can't imagine rational people, presented with the evidence that homosexuality is biological and not a choice, would continue to discriminate."[59] The logic of a developing genetic essentialism seems to be this: if male homosexuality is genetically inherited, then those with this inheritance can claim rights based on this inheritance. If sexual orientation is genetically determined and not a choice, then gay men can claim the ethical high ground to justify any sort of gay lifestyle. Yet, I ask, might such ethical logic be buying too quickly into the gene myth? Would it be worth our time to pause and project the various directions that ethical reasoning might take us? Are there other possible directions?

I would suggest here that if we eventually accept as fact that male homosexual orientation is genetically inherited, then the ethical logic that follows could go a number of directions. *The scientific fact does not itself determine the direction of the ethical interpretation of that fact.* To demonstrate this, I would like to begin with a basic question: Does the genetic predisposition toward homosexuality make the bearer of that gene innocent or guilty? Two answers are logically possible. On the one hand, one could claim that because a homosexual man inherited the gay gene and did not choose a gay orientation by his own free will, he is innocent. On the other hand, one could take the opposite road and identify the gay gene with the predisposition to sexual sin. Society could claim that the body inherited by each of us belongs to who we are—who I am as a self is determined at least in part by what my parents bequeathed me—and that an inherited disposition to homosexual behavior is just like other innate dispositions such as lust or greed or similar forms of concupiscence that are shared with the human race generally; and all this constitutes the state of original sin into which we are born. Signposts point in both ethical directions.

The science and the ethical reflection on the science here may jolt some theologians into remembering that nearly forgotten doctrine, namely, original sin. Setting aside for a moment our judgment as to whether homosexual behavior is moral or immoral,[60] let us ask: Does our biological predisposition toward a specific behavior in itself make that behavior moral or immoral? If it is natural, is it automatically good? Not necessarily, according to the concept of original sin. To my observation, religious groups and their theologians have not yet placed this issue on their agenda.

Yet the current scientific situation—actually, the current culture with its gene myth picturing science as deterministic—is compelling us to open up once again the case for original sin. The popularity of Richard Dawkins's book *The Selfish Gene*, along with the controversy created by sociobiologists, demonstrates a growing interest in the prospect that scientists will be able to explain more and more of human behavior in biological terms.[61] Edward O. Wilson defines sociobiology as "the systematic study of the biological basis of all social behavior"; and he has staked out the biological claim rather forcefully: "The genes hold culture on a leash. The leash is very long, but inevitably values will be constrained in accordance with their effects on the human gene pool."[62] The point here is that the debate over the gay gene is taking place within a wider cultural debate—the nature vs. nurture debate—regarding the primary influence on human belief and behavior. "In the long debate over the relative influences of nature and nurture," write Dorothy Nelkin and Laurence Tancredi, "the balance seems to have shifted to the biological extreme."[63]

With nature winning over nurture in the gene myth, we will be tempted to ground our morality in nature. But will this be as easy as it looks? What if we were to discover a genetic predisposition to homophobia that leads to gay bashing? Would both the gay gene and the homophobia gene be equally moral? Equally immoral? Would we not have to appeal to extrabiological or extrascientific criteria to adjudicate?

According to the concept of original sin, our nature is fallen. As fallen, our nature can lead us toward sin. Our natural inclinations are not by definition good. What is good is determined by the leadings of God, not the leadings of our biology in any strict sense. Yet the concept of original sin cannot by any means be reduced to a natural—read genetic—propensity to do evil. Much more is going on here.

Let us refer to Saint Augustine for a moment. This North African bishop affirmed that God is good and that nature is created good; however, he believed that sin is something we can and should fully expect for the foreseeable future. Sin is naturally unnecessary, yet historically inevitable.[64] Why? What we have since come to know as the doctrine of original sin is Augustine's answer. It goes like this. By nature and at an early point in history God created the first human beings good. We know our original parents as Adam and Eve. These two committed the first sin. In Lamarckian fashion, this historical incident affected their biological nature so that the propensity for sinning became henceforth an inheritable trait. We inherit the predisposition to sin, even though our sins are our own acts. Augustine likened original sin to a disease that we inherit congenitally and pass on to our children. Adam's original sin becomes for us inherited sin, a disposition that is not necessary by nature but for us is historically inevitable. In terms of world history, sin is the historical corruption of an originally good nature.

Augustine was led to this explanation because of a fundamental commitment to the unity of the human race. This is a double unity, a unity of sin and a unity of salvation. He was taking with utmost seriousness passages in the writings of Paul such as 1 Corinthians 15:22: "As all die in Adam, so all will be made alive in Christ" (see also Rom. 5:12–16). Your and my sins today do not merely imitate Adam's sin; somehow they participate and mutually penetrate and continue to propagate. Augustine was not primarily concerned with the specific propensities each of us has for lust or envy or whatever. That was secondary. Primary was the sense that all of us in the human race are in the same boat, the same sinful boat that is sailing away from consciousness of God and away from love for one another. This unity in sin is the correlate of our unity in salvation. Augustine wrote, "We have derived from Adam, in whom all have sinned, not all our actual sins, but only original sin; whereas from Christ, in whom we are all justified, we obtain the remission not merely of that original sin, but of the rest of our sins also, which we have added."[65] For Augustine, the Savior, Jesus Christ, is an individual human being, to be sure; yet he is much more. Christ is the prototypical human being, the eternal Logos, and the image of the Divine—the true *imago Dei*—under the conditions of humanity. The whole human race finds its definition, its identity, and its rescue from inherited sin through the forgiving and resurrecting power of Christ.

If we try to apply the Augustinian notion of original sin to the present discussion, it would seem that the gay gene would find its place in the larger description of a human condition that includes all of us. The homosexual predisposition and perhaps even a genetic predisposition to homophobia and gay bashing would together constitute signs of a fallen human nature, a historically specific form of human nature in which desire threatens to cause psychic and social strife. The responsible or Augustinian road would ask the bearer of the gay gene to admit what all human beings are asked to admit, namely, we have been born with an inherited predisposition toward sin, and our ethical mandate is to strive for a life of love that transcends our inborn desires. Whether a given individual has the gay gene or the homophobic gene, the ethical mandate to love one another applies to both.

One of the major weaknesses in the Augustinian solution is that we in the modern world cannot split nature and history in quite the same way he did. Augustine could begin with nature created good and then identify a point in historical time—the time of Adam and Eve in the Garden of Eden—to locate the originating sin that corrupted the nature that today we inherit in our genes. There are two reasons why this is unacceptable today. First, the current scientific worldview informed by evolutionary history makes no room in its time scheme for an Eden story—that is, we cannot go back in time to find a point prior to which nature was benign and after which it had fallen. Second, evolutionary theory and such things as the big bang theory in physical cosmology are leading scientists to view nature itself as history. Nature itself is not fixed or eternal, but subject to contingency and change. So, bifurcating reality into two discrete realms, nature and history, no longer makes sense.

In sum, the Christian understanding of original sin as bequeathed from Saint Augustine has less to do with biological determinants of our behavior and more to do with the unity all we humans share with one another both in Adam and in Christ. Nevertheless, this theological tradition will be skeptical of arguments that seek moral approval on the basis of genetic determinism. The gene myth has no automatic theological endorsement. To reiterate: the scientific fact does not itself determine the direction of the ethical interpretation of that fact.

In the meantime, what should theologians do about laboratory discoveries such as that of the gay gene? Meeting in Berkeley, California, on August 17,

1993, the geneticists, ethicists, and theologians working together on a research project conducted by the Center for Theology and the Natural Sciences at the Graduate Theological Union dealing with "Theological and Ethical Questions Raised by the Human Genome Initiative" framed the following statement:

> The recent report of an X-linked predisposition to some forms of male homosexuality has already created a furor, and we think it appropriate to pause and reflect critically if briefly upon what implications may or may not be justified. Although many critics have emphasized the high quality of the researchers and of the design of the research, they also point out that the results are still preliminary. Assuming that the research is replicated and supported by additional experimental designs, the scientific fact still does not itself determine the direction of the ethical interpretation of that fact. The ethical logic could go in several directions. On the one hand, genetic predisposition may remove sexual preference from the list of behaviors that can be blamed on individual choice or on the behavior of parents. On the other hand, it may lead to a cry to develop "cures" or eugenic measures to eliminate the predispositional gene, as have earlier studies of diversity between races lent misplaced credibility to discriminatory practices.
>
> The possible genetic predisposition to some forms of homosexuality is but one instance of a broader theme of naturally occurring human genetic diversity. The problem of how to recast notions of freedom and responsibility in light of that diversity remains a basic philosophical issue. The honest expression of diverse opinions within the lesbian and gay community to the recently publicized claims shows the kind of maturity that we appreciate and commend to other social commentators. As theological and scientific commentators, we urge that the best of genetic research be taken seriously. There is a rich variety of theological reflection both in the religious traditions and in recent developments which leads us to affirm that the theological questions raised by this genetic research are profound and subtle. We recognize that genetic diversity requires a response of love, respect and justice.[66]

7. Somatic vs. Germ-Line Intervention

A seventh ethical issue that appears in both secular and religious discussions is the distinction between two pairs of concerns: somatic intervention vs. germ-line intervention and therapy vs. enhancement. By "somatic therapy" we refer to the treatment of a disease in the body cells of a living individual by trying to repair an existing defect. Body cells contain the full complement of forty-six chromosomes. In contrast, germ-line cells in the reproductive process contain only twenty-three chromosomes, and these are passed on to the next generation. Intervention into the germ-line cells would influence heredity; and the possibility of significantly enhancing the quality of life of future generations through germ-line intervention is looming on the genetic horizon.[67]

All ethical commentators I have reviewed agree that somatic therapy is morally desirable, and they look forward to the advances HGP will bring for expanding this important work. Yet the ethically minded stop short of endorsing genetic selection and manipulation for the purposes of enhancing the quality of biological life for otherwise normal individuals or for the human race as a whole. We can speculate that the new knowledge gained from HGP might locate genes that affect the brain's organization and structure so that careful engineering might lead to enhanced ability for abstract thinking or to other forms of physiological and mental improvement.[68] But such speculations are greeted with the greatest caution. Molecular hematologist W. French Anderson says,

> Somatic cell gene therapy for the treatment of severe disease is considered ethical because it can be supported by the fundamental moral principle of beneficence: It would relieve human suffering. Gene therapy would be, therefore, a moral good. Under what circumstances would human genetic engineering not be a moral good? In the broadest sense, when it detracts from, rather than contributes to, the dignity of man. . . . Somatic cell enhancement engineering would threaten important human values in two ways: It could be medically hazardous. . . . And it would be morally precarious, in that it would require moral decisions our society is not now prepared to make, and it could lead to an increase in inequality and discriminatory practices.[69]

In short, genetic enhancement risks violating human dignity by opening up the possibility of discrimination.

Religious ethicists agree: somatic therapy should be pursued, but enhancement especially through germ-line engineering raises cautions about protecting human dignity. The World Council of Churches is representative. In a 1982 document we find these comments:

> Somatic cell therapy may provide a good; however, other issues are raised if it also brings about a change in germ-line cells. The introduction of genes into the germ-line is a permanent alteration. . . . Nonetheless, changes in genes that avoid the occurrence of disease are not necessarily made illicit merely because those changes also alter the genetic inheritance of future generations. . . . There is no absolute distinction between eliminating "defects" and "improving" heredity.[70]

The primary caution raised by the WCC here has to do with our lack of knowledge regarding the possible consequences of altering the human germ line. The problem is that the present generation lacks sufficient information regarding the long-term consequences of a decision made today that might turn out to be irreversible tomorrow. Thus, the WCC does not forbid forever germ-line therapy or even enhancement; rather, it cautions us to wait and see. Similarly, the Methodists "support human gene therapies that produce changes that cannot be passed on to offspring (somatic), but believe that they should be limited to the alleviation of suffering caused by disease."[71] The United Church of Christ also approves "altering cells in the human body, if the alteration is not passed to offspring."[72]

The Catholic Health Association is more positive: "Germ-line intervention is potentially the only means of treating genetic diseases that do their damage early in embryonic development, for which somatic cell therapy would be ineffective. Although still a long way off, developments in molecular genetics suggest that this is a goal toward which biomedicine could reasonably devote its efforts."[73]

Another reason for caution regarding germ-line enhancement, especially among the Protestants, is the specter of eugenics. The word "eugenics" connotes the ghastly racial policies of Nazism, and this accounts for much of today's mistrust of genetic science in Germany and elsewhere.[74] In 1992, Germany passed legislation prohibiting germ-line intervention. No one expects a resurrection of the Nazi nightmare; yet some critics fear a subtle form of eugenics slipping in the cultural back door.[75]

The growing power to control the design of living tissue will foster the emergence of the image of the "perfect child," and a new social value of perfection will begin to oppress all those who fall short. Although the perfect child syndrome is not yet widely discussed in the published literature, religious ethicists speaking in March 1992 at the Genetics, Religion and Ethics Conference held at the Texas Medical Center in Houston saw the image of the "perfect child" to be a clear and present danger.[76]

The Methodists categorically oppose gene therapy "for eugenic purposes or genetic enhancement designed merely for cosmetic purposes or social advantage."[77] The National Council of Churches takes a cautious but calm attitude. Even though eugenic programs can be envisioned that would seek total eradication of particular recessive genetic disorders (let alone enhancements), such programs would require mandatory prenatal diagnosis of at-risk pregnancies and other procedures on an impossibly large scale. There is no sign that such far-reaching programs could be implemented. In addition, while the NCC recognizes how past efforts to improve heredity have been associated with prejudices and racial chauvinism, it recommends that we suspend judgment regarding current genetic research until we know more: "It is important today to separate out common misconceptions of the past and present from valid knowledge about heredity."[78]

Where does the debate among bioethicists stand? Eric T. Juengst summarizes five arguments in favor of germ-line modification for the purposes of therapy:

1. *Medical utility:* germ-line gene therapy offers a true cure for many genetic diseases.
2. *Medical necessity:* such therapy is the only effective way to address some diseases.
3. *Prophylactic efficiency:* prevention is less costly and less risky than cure.
4. *Respect for parental autonomy* when parents request germ-line intervention.
5. *Scientific freedom* to engage in germ-line inquiry.

Juengst also summarizes five arguments opposing germ-line intervention:

1. *Scientific uncertainty* and risks to future generations.
2. *Slippery slope to enhancement* that could exacerbate social discrimination.
3. *Consent of future generations* is impossible to get.
4. *Allocation of resources:* germ-line therapy may never be cost effective.
5. *Integrity of genetic patrimony:* future generations have the right to inherit a genetic endowment that has not been intentionally modified.[79]

This is where we stand.

8. Are We Asking Our Scientists to Play God?

The debate over germ-line intervention brings us directly to the question popularized by newspaper headlines: Should we ask our scientists to play God? Should we ask them to refrain from playing God? The way the question is posed in the press is usually so superficial as to be misleading. Yet beneath the superficiality we find a theological issue of consequence, namely, do we as human beings share with our creator God some responsibility for the ongoing creativity of our world? Here we will ask just what is meant by the phrase "play God" and then turn to the connection between divine and human creativity. The concept of creation includes anthropology and the notion that the human race is created in the divine image.[80] In light of what I said earlier regarding the promise of new creation, I would like to go beyond reporting what religious communities are saying and offer my theological argument: if we understand God's creative activity as giving the world a future, and if we understand the human being as a created co-creator, then ethics begins with envisioning a better future. This suggests we should at minimum keep the door open to improving the human genetic lot. The derisive use of the phrase "play God" should not deter us from shouldering our responsibility for the future. To seek a better future is to "play human" as God intends us to do.[81]

The acerbic rhetoric that usually employs the phrase "play God" is aimed at inhibiting, if not shutting down, certain forms of scientific research and medical therapy. This applies particularly to the field of human genetics and, still more particularly, to the prospect of germ-line intervention for purposes of human enhancement—that is, the insertion of new gene segments of DNA into sperm or eggs before fertilization or into undifferentiated cells of an early embryo that will be passed on to future generations and may become part of the permanent gene pool. Some scientists and religious spokespersons are putting a chain across the gate to germ-line intervention with a posted sign reading, "Thou shalt not play God." A *Time*/CNN poll cites a substantial majority (58 percent) who believe altering human genes is against the will of God.[82]

The question about "playing God" leads to correlate questions. Does our genetic makeup represent a divine creation in such a way that it is complete and final as it is? Is our DNA sacred? Are we desecrating a sacred realm when we try to discern the mysteries of DNA, and are we exhibiting excessive human pride when we try to engineer our genetic future?

In light of these questions the curious phrase "playing God" can have at least three overlapping meanings. The first and somewhat benign meaning refers to the sense of awe rising up from the new discoveries into the depths of life. Science and its accompanying technology are drawing light down into the hitherto dark and secretive caverns of human reality. Mysteries are becoming revealed; and the scientists who are the revealers sense that we are on the threshold of acquiring godlike powers. At this level we do not yet have any reason to object to research. Rather, what we have here is a welcome expression of awe.[83]

The second meaning of "playing God" refers to the clinical setting, where medical doctors such as surgeons appear to have life-and-death powers over their patients. Neither of these two meanings is particularly relevant to the genetic issues at hand.

The third meaning of "playing God" is most relevant to genetic science. It refers to placing ourselves where God and only God belongs. In a 1980 letter warning then President Jimmy Carter, Roman Catholic, Protestant, and Jewish spokespersons used the phrase "playing God" to refer to individuals or groups who would seek to control life-forms. Any attempt to "correct" our mental and social structures by genetic means to fit one group's vision of humanity is dangerous.[84]

Why do such critics of genetic research prescribe a new commandment,

"Thou shalt not play God"? The answer here is this: because human pride or hubris is dangerous. We have learned from experience that what the Bible says is true: "Pride goes before destruction" (Prov. 16:18). And in our modern era pride among the natural scientists has taken the form of overestimating our knowledge, of arrogating for science a kind of omniscience that we do not have. Or to refine it a bit, "playing God" means we confuse the knowledge we do have with the wisdom to decide how to use it. Frequently lacking wisdom we falsely assume we possess, scientific knowledge leads to unforeseen consequences such as the potential destruction of the ecosphere. Applied to genetic therapy, the commandment against "playing God" implies that the unpredictability of destructive effects on the human gene pool should lead to a proscription against germ-line intervention. In light of this, "there is general agreement that human germ-line intervention for any purpose should always be governed by stringent criteria for safety and predictability."[85] This "general agreement" seeks to draw upon wisdom to mitigate pride.

A related implication of the phrase "playing God" is that DNA has come to function, in effect, as an inviolable sacred, a special province of the Divine that should be off limits to mere mortals. Robert Sinsheimer, among others, suggests that when we see ourselves as the creators of life, then we lose reverence for life.[86] This lack of reverence for life as nature has bequeathed it to us drives Jeremy Rifkin to attack the kind of genetic research that will lead to algeny—that is, to "the upgrading of existing organisms and the design of wholly new ones with the intent of 'perfecting' their performance." The problem with algeny is that it represents excessive human pride. "It is humanity's attempt to give metaphysical meaning to its emerging technological relationship with nature."[87] Rifkin's message is: let nature be! Don't try to make it better! In advocating this hands-off policy, Rifkin does not appeal to Christian or Jewish or other theological principles. Rather, he appeals to a vague naturalism, according to which nature itself claims sacred status. He issues his own missionary's call: "The resacralization of nature stands before us as the great mission of the coming age."[88]

What is the warrant for treating nature in general, or DNA specifically, as sacred and therefore morally immune from technological intervention? Ronald Cole-Turner criticizes Sinsheimer and Rifkin for making an unwarranted philosophical and theological leap from the association of DNA with life to the metaphysical proscription against technical manipulation.

Is DNA the essence of life? Is it any more arrogant or sacrilegious to cut DNA than to cut living tissue, as in surgery? It is hard to imagine a scientific or philosophical argument that would successfully support the metaphysical or moral uniqueness of DNA. Even DNA's capacity to replicate does not elevate this molecule to a higher metaphysical or moral level. Replication and sexual reproduction are important capacities, crucial in biology. But they are hardly the stuff of sanctity.[89]

To nominate DNA for election into the halls of functional sacrality, says Cole-Turner, is arbitrary. Theologians in particular should avoid this pitfall. "To think of genetic material as the exclusive realm of divine grace and creativity is to reduce God to the level of restriction enzymes, viruses, and sexual reproduction. Treating DNA as matter—complicated, awe-inspiring, and elaborately coded, but matter nonetheless—is not in itself sacrilegious."[90]

What the three meanings of "playing God" raise up for us is the question of the relationship between the Divine Creator and the human creature. We can distinguish Christian and Jewish theism from a naturalism that reveres life. Theists hold that God is the creator of all things, including life. Natural life, important as it is, is not ultimate. God is ultimate.

One can argue to this position on the basis of *creatio ex nihilo*, creation out of nothing. All that exists has been called from nothing by the voice of God and brought into existence, and at any moment could in principle return to the nonexistence from which it came. Life, as everything else in existence, is finite, temporal, and mortal. The natural world depends upon a Divine Creator who transcends it. Nature is not its own author. Nor can it claim ultimacy, sanctity, or any other status rivaling God. This leads biologist Hessel Bouma III and his colleagues at the Calvin Center for Christian Scholarship to a pithy proposition: "God is the creator. Therefore, nothing that God made is god, and all that God made is good." This implies, among other things, that we should be careful when accusing physicians and scientists of "playing God." We must avoid idolatrous expectations of technology, to be sure; "but to presume that human technological intervention violates God's rule is to worship Mother Nature, not the creator. Natural processes are not sacrosanct."[91]

One can also argue to this position on the basis of *creatio continua*, continuous creation—that is, on the basis of the idea that creation is ongoing, one can argue for human intervention and contribution to the process. God did not just extricate the world from the divine assembly line like a car, fill its tank with gas, and then let it drive itself down the highways of history. Divine steering, braking, and accelerating still go on. The creative act whereby God brought the world into existence *ab initio*, at the beginning, is complemented with God's continued exercise of creative power through the course of natural and human history. The God of the Bible is by no means absent. This God enters the course of events, makes promises, and then fulfills them. God is the source of the new. Just as the world appeared new at the beginning, God continues to impart the new to the world and promises a yet outstanding new creation to come.

My way of conceiving of *creatio ex nihilo* together with *creatio continua* is this way: the first thing God did was to give the world a future.[92] The act of drawing the world into existence from nothing is the act of giving the world a future. As long we have a future, we exist. When we lose our future, we cease to exist. God continues moment by moment to bestow futurity, and this establishes continuity while opening reality up to newness. Future giving is the way in which God is creative. It is also the way God redeems. God's grace comes to the creation through creative and redemptive future giving.

God creates new things. The biblical description of divine activity in the world includes promises and fulfillments of promises. This implies two divine qualities. First, God is not restricted to the old, not confined by the status quo. God may promise new realities and then bring them to pass. The most important of the still outstanding divine promises is that of the "new creation" yet to come. Second, this God is faithful, trustworthy. On the basis of the past record, the God of Israel can be trusted to keep a promise. We can trust God's creative and redemptive activity to continue in the future.

The next step in the argument is to conceive of the human being as the created co-creator. The term "created co-creator" comes from the work of Philip Hefner.[93] The term does a couple of important things. First, the term "created" reminds us that the way God creates differs from the way we human beings create. God creates *ex nihilo*. We have been created by God. We are creatures. So, whatever creativity we manifest cannot share rank with creation out of nothing, on the same level with our Creator. Yet, second, the term "co-creator" signifies what we all know, namely, that creation does not stand still. It moves. It changes. So do we. Furthermore, we have partial influence on the direction it moves and the kind of changes that take place. We are creative in the transformatory sense. Might we then think of the *imago Dei*—the image of God embedded in the human race—in terms of creativity? Might we think of ourselves as co-creators, sharing in the transforming work of God's ongoing creation?

Human creativity is ambiguous. We are condemned to be creative. We cannot avoid it. The human being is a tool maker and a tool user. We are *homo faber*. We cannot be human without being technological, and technology changes things for good or ill. Technology is usually designed for good reasons such as service to human health and welfare, but we know all too well how shortsightedness in technological advances does damage. This is indirect evil. Direct evil is also possible. Technology can be pressed into the service of violence and war, as in the making of weapons. It is by no means an unmitigated good. Yet despite its occasional deleterious consequences, we humans have no choice but to continue to express ourselves technologically and, hence, creatively.

We cannot *not* be creative. The ethical mandate, then, has to do with the purposes toward which our creativity is directed and the degree of zeal with which we approach our creative tasks.

Conclusion: Theological Commitments to Human Dignity

The ethical issues just described are but a small sample drawn from a longer list that would include questions regarding the social implications of defining a "defective" gene, equity of access to genetic services, gender justice, environmental impact, patenting knowledge of DNA sequences or new life-forms, developing biological weapons, eugenics, and numerous others. Some of our farsighted religious leaders have entered into serious conversation with conscientious scientists so that cooperative thinking about our response and responsibility for the future can be anticipated.[94] "Science and studies in genetics and biology have given us a fascinating insight into our world," writes ethicist Thomas Shannon, and only through a careful examination of this insight "can we be responsible to ourselves, to our ancestors, and to our environment."[95]

Virtually all Roman Catholics and Protestants who take up the challenge of the new genetic knowledge seem to agree on a handful of theological axioms. First, they affirm that God is the creator of the world and, further, that God's creative work is ongoing.[96] God continues to create in and through natural genetic selection and even through human intervention in the natural processes. Second, the human race is created in God's image. In this context, the divine image in humanity is tied to creativity. God creates. So do we. With surprising frequency, we humans are described by theologians as "co-creators" with God, making our contribution to the evolutionary process.[97] To avoid the arrogance of thinking that we humans are equal to the God who created us in the first place, we must add the term "created" to make the phrase "created co-creators." This emphasizes our dependency upon God while pointing to our human opportunity and responsibility. Third, these religious axioms place a high value on human dignity.

By "dignity" I mean what philosopher Immanuel Kant meant, that we treat each human being as an end, not merely as a means to some further end. Although this may be a minimal definition, Christian churches mean this and more by human dignity. The United Church of Canada eloquently voices the dominant view: "In non-theological terms it [dignity] means that every human being is a person of ultimate worth, to be treated always as an end and not as a means to someone else's ends. When we acknowledge and live by that principle our relationship to all others changes."[98] Pope John Paul II begins to appropriate the dignity principle in an elocution where he condemns "in the most explicit and formal way experimental manipulations of the human embryo, since the human being, from conception to death, cannot be exploited for any purpose whatsoever."[99] As church leaders respond responsibly to new developments in HGP, we can confidently forecast one thing: this affirmation of dignity will become decisive for thinking through the ethical implications of genetic engineering. Promoting dignity is a way of drawing an ethical implication from what the theologian can safely say, namely, that God loves each human being regardless of our genetic makeup and, therefore, we should love one another according to this model.

Yet there is more. The theology of co-creation leads Ronald Cole-Turner to a beneficent vision: "For the church, it is not enough to avoid the risks. Genetic engineering must contribute in a positive way to make the world more just and more ecologically sustainable, and it must contribute to the health and nutrition of all humanity."[100]

Notes

1. This chapter synthesizes and revises research drawn from numerous directions, and some of the material presented here has appeared in partial form in journals such as the *Journal of Medicine and Philosophy, Forum for Applied Research and Public Policy, Midwest Medical Ethics, Dialog, Christian Century,* in chap. 10, "Genes and Sin," of my book *Sin: Radical Evil in Soul and Society* (Grand Rapids, Mich.: Eerdmans, 1994), and in my book *Playing God? Genetic Determinism and Human Freedom* (New York and London: Routledge, 1997).

2. HGP has been criticized because (1) as "big science" with top-down administration, it seems to squeeze out local initiative by smaller laboratories, and (2) the large financial investment in genome mapping and sequencing leaves little funds left over for other worthy research projects. "The genome project has been overhyped and oversold," complains Boston University School of Medicine health law professor George J. Annas. He then adds, "It is the obligation of those who take legal and ethical issues seriously to insure that the dangers, as well as the opportunities, are rigorously and publicly explored." "Who's Afraid of the Human Genome?" *Hastings Center Report* 19, no. 4 (July–August 1989): 21.

3. At this writing, Daniel Cohen at the Center for the Study of Human Polymorphism in Paris has just produced the first map, but it is yet sketchy and riddled with errors; and Craig Venter at the Institute for Genomic Research near Washington, D.C., plans to identify all genes in the near future. As of 1996, the best map has sixteen thousand markers. The full sequencing of nucleotides will take much longer.

4. James D. Watson contends that "the more important reward is satisfying one's curiosity about how nature operates, and for biologists this means a deeper

understanding of the nature of living organisms." "The Human Genome Project: Past, Present, and Future," *Science* 248 (April 6, 1990): 48.

5. Ibid., 46.

6. Healy instituted a plan to seek patents for NIH-funded discoveries so as to head off patents by private biotech companies. Watson objected, arguing that the secrecy surrounding the patent process would tend to restrict the free flow of scientific information necessary to facilitate research.

7. See Ann Lammers and Ted Peters, "Genethics: Implications of the Human Genome Project," *Christian Century* 107, no. 27 (October 3, 1990): 868–72; *Genethics: The Clash Between the New Genetics and Human Values,* ed. David Suzuki and Peter Knudtson, rev. ed. (Cambridge: Harvard University Press, 1990); and David Heyd, *Genethics: Moral Issues in the Creation of People* (Berkeley: University of California Press, 1992).

8. "The task of the theologians is to bring their discipline to bear on the fundamental questions posed by ethics," writes Trutz Rendtorff, *Ethics,* 2 vols. (Minneapolis: Fortress Press, 1986–89), 1:10.

9. My focus here will be on recent work among Roman Catholic and Protestant spokespersons. Elsewhere in this volume we will look specifically at the Jewish community. Though beyond the scope of this book, it would be well to elicit the perspectives of other religious groups in the near future. In preliminary fashion we note that for Islam, "there are no restrictions on genetic research, but that the biotechnological applications of such research should be individually allowed through juridical sanctions that ensure their use for the benefit of humanity." Hassan Hathout and B. Andrew Lustig, "Bioethical Developments in Islam," in *Theological Developments in Bioethics: 1990–1993,* vol. 3 of *Bioethics Yearbook,* ed. B. Andrew Lustig (Boston: Kluwer, 1993), 145. Buddhists exact the standard of compassion "to avoid the eugenic tendencies associated with the Western predilection for attempting to create ideal situations." Kathleen Nolan, "Buddhism, Zen, and Bioethics," in ibid., 207.

10. Office of Family Ministry and Human Sexuality (Office 711, 475 Riverside Drive, New York, NY 10027), *Human Life and the New Genetics: A Report of a Task Force Commissioned by the National Council of Churches of Christ in the U.S.A.,* 1980, p. 47 (hereinafter: NCC).

11. The Church of the Brethren, "Statement on Genetic Engineering," *1987 Annual Conference Minutes,* 455.

12. "Genetic discrimination is defined as discrimination against an individual or against members of that individual's family solely because of real or perceived differences from the normal genome of that individual." Paul R. Billings, Mel A. Kohn, Margaret de Cuevas, Jonathan Beckwith, Joseph S. Alper, and Marvin R. Natowicz, "Discrimination as a Consequence of Genetic Testing," *American Journal of Human Genetics* 50 (1992): 481.

13. Eliot Marshall, "A Tough Line on Genetic Screening," *Science* 262, no. 5136 (November 12, 1993): 985. Some members of the IOM insist that screening

should continue for two diseases—phenylketonuria (PKU) and hypothyroidism—because they can be treated directly.

14. See Thomas H. Murray, "Ethical Issues in Human Genome Research," *FASEB Journal* 5 (January 1991): 55–60.

15. Mitchel L. Zoler, "Genetic Tests," *Medical World News*, January 1991, 1–4. "It is clear that unfair and discriminatory uses of genetic data already occur under current conditions. Enacted state and federal laws are inadequate to prevent some forms of genetic discrimination particularly that due to the health insurance industry." Billings et al., "Discrimination as a Consequence of Genetic Testing," 481. A *Time*/CNN poll on December 2, 1993, asked, "Do you think it should be legal for employers to use genetic tests in deciding whom to hire?" The responses: 87 percent answered no; 9 percent answered yes. Philip Elmer-DeWitt, "The Genetic Revolution," *Time*, January 17, 1994, 46–53.

16. Alexander Morgan Capron, "Which Ills to Bear? Reevaluating the 'Threat' of Modern Genetics," *Emory Law Journal* 39, no. 3 (summer 1990): 690.

17. Reporting on the Working Group on Ethical, Legal, and Social Issues (ELSI) jointly sponsored by the National Institutes of Health (NIH) and the Department of Energy (DOE), Elinor J. Langfelder and Eric T. Juengst write, "The Americans with Disabilities Act prohibits discriminatory actions based on a person's genotype, including the possibility of having affected children," and argue that preemployment medical tests should be limited to those necessary to assess job-related physical and mental conditions. "Social Policy Issues in Genome Research," *Forum for Applied Research and Public Policy* 8, no. 3 (fall 1993): 15. The insurance industry tends to oppose restrictions in favor of regulations. "Genetic testing restrictions may not adequately protect a person's access to insurance," argues Ami S. Jaeger, because current use of medical histories already yields considerable genetic information. It is better to have the genetic information and to use it "within a regulatory framework that provides universal access and adequately protects the rights of individuals." "An Insurance View on Genetic Testing," *Forum for Applied Research and Public Policy* 8, no. 3 (fall 1993): 25. As of this writing eleven states now prohibit discrimination in health insurance, seven states prohibit discrimination in employment, and four states prohibit genetic discrimination in life insurance. See Michael S. Yesley, "Genetic Privacy, Discrimination, and Social Policy: Challenges and Dilemmas," *Microbial and Comparative Genomics* 2, no. 1 (1997): 20.

18. Invoking the principle of confidentiality in order to avoid job discrimination is seriously considered by the Joint NIH-DOE Working Group on Ethical, Legal, and Social Issues (ELSI) of human genome research. Yet, more important, ELSI supports "universal access to . . . basic health services" and states that "information about past, present or future health status, including genetic information, should not be used to deny health care coverage or services to anyone." Rather than rely simply on protection of privacy, the point here is

that genetic information, whether available or not available, should not affect access to basic health care. National Center for Human Genome Research, *Genetic Information and Health Insurance,* Publication no. 93-3686 (Washington, D.C.: National Institutes of Health, 1993), 2.

19. *Biotechnology: Its Challenges to the Churches and the World,* Report by World Council of Churches Subunit on Church and Society (Geneva: WCC, 1989), 2 (hereinafter: WCC).

20. "United Methodist Church Genetic Task Force Report to the 1992 General Conference," 118 (hereinafter: Methodist).

21. "The Church and Genetic Engineering," Pronouncement of the Seventeenth General Synod, United Church of Christ, Fort Worth, Texas, 1989, 4 (hereinafter: UCC). See the interpretation and commentary of this pronouncement by Ron Cole-Turner, "Genetics and the Church," *Prism* 6, no. 1 (spring 1991): 53–61.

22. In December 1993, China announced a program of abortions, forced sterilization, and marriage restrictions to "avoid new births of inferior quality and heighten the standards of the Chinese people." See *Time,* January 17, 1994, 53. This is eugenics old-fashioned style as practiced earlier in the twentieth century by England, the United States, and Nazi Germany. The genetics revolution may make eugenics more efficient; and as more and more reproductive clinics open for business, we may see a version of free market eugenics.

23. John A. Robertson, "Procreative Liberty and Human Genetics," *Emory Law Journal* 39, no. 3 (summer 1990): 707.

24. *Gaudium et Spes,* in *Catholic Social Thought: The Documentary Heritage,* ed. David O'Brien and Thomas A. Shannon (Maryknoll, N.Y.: Orbis Books, 1992), 199.

25. Congregation for the Doctrine of the Faith, "Instruction on Respect for Human Life in Its Origin and on the Dignity of Procreation: Replies to Certain Questions of the Day," in *Bioethics,* ed. Thomas A. Shannon, 3d ed. (New York: Paulist Press, 1987), 599.

26. Pope John Paul II, *The Gospel of Life* (New York: Random House, Times Books, 1995), 25.

27. Catholic Health Association of the United States, *Human Genetics: Ethical Issues in Genetic Testing, Counseling, and Therapy* (St. Louis: Catholic Health Association, 1990), 21.

28. "A Social Statement on Abortion," adopted by the second biennial Churchwide Assembly of the Evangelical Lutheran Church in America (ELCA) meeting in Orlando, Florida, August 28–September 4, 1991 (hereinafter: Lutheran). As of this writing an ELCA task force is beginning a study on genetic testing and screening.

29. WCC, 31–32.

30. The Division of Mission in Canada, "A Brief to the Royal Commission on

New Reproductive Technologies on Behalf of the United Church of Canada," January 17, 1991, 13–14 (hereinafter: United Church of Canada).

31. UCC, 4.

32. Cole-Turner, "Genetics and the Church," 57.

33. Cited by Christopher Anderson, "NIH Drops Bid for Gene Patents," *Science* 263 (February 18, 1994): 909.

34. Venter and his colleague, William A. Haseltine, CEO of Human Genome Sciences, made the cover of *Business Week*, May 8, 1995, with an article by John Carey, "The Gene Kings," 72–78.

35. James D. Watson, "A Personal View of the Project," in *The Code of Codes: Scientific and Social Issues in the Human Genome Project*, ed. Daniel J. Kevles and Leroy Hood (Cambridge: Harvard University Press, 1992), 164–73.

36. Cited in L. Roberts, "NIH Gene Patents, Round Two," *Science* 255 (1992): 912.

37. P. Zurer, "Critics Take Aim at NIH's Gene Patenting Strategy," *Chemical and Engineering News* 70 (1992): 21–22.

38. C. Anderson, "To Patent a Naked Gene," *Nature* 353 (1991): 485. Rebecca S. Eisenberg says that the argument against allowing NIH to patent Venter's sequences is not really that these sequences are useless; rather, it is that we do not yet know what they are good for. We should not be able to claim patent rights ahead of subsequent researchers who will attempt to figure this out. "Genes, Patents, and Product Development," *Science* 257 (August 14, 1992): 905.

39. 33 U.S. 130 (1948); cited by Eisenberg, "Genes, Patents, and Product Development," 904.

40. Kevles and Hood, *The Code of Codes*, 313, 314.

41. "HUGO Statement on Patenting of DNA Sequences" (HUGO Americas, 7986-D Old Georgetown Road, Bethesda, MD 20814), report in *Human Genome News* 6, no. 6 (March-April 1995): 5.

42. Andrew Kimbrell, *The Human Body Shop* (San Francisco: Harper, 1993), 190.

43. General Board of Church and Society of the United Methodist Church, 100 Maryland Avenue, NE, Washington, D.C. 20002.

44. Battle imagery was used in the front page coverage of the *New York Times*, national edition, May 13, 1995, "Religious Leaders Prepare to Fight Patents on Genes," 1.

45. Jeremy Rifkin, *Algeny* (New York: Viking, 1983), 252. See Gary Stix, "Profile: Jeremy Rifkin," *Scientific American* 277, no. 2 (August 1997): 28–32.

46. John Maddox, "Has Nature Overwhelmed Nurture?" *Nature* 366 (November 11, 1993): 107.

47. R. David Cole, "Genetic Predestination?" *Dialog* 33, no. 1 (winter 1994): 21.

48. Walter Gilbert, "Towards a Paradigm Shift in Biology," *Nature* 349 (1991): 99.

49. L. Jaroff, "The Gene Hunt," *Time*, March 20, 1989, 67.

50. Jonathan Beckwith, "A Historical View of Social Responsibility in Genetics," *BioScience* 43, no. 5 (May 1993): 192.

51. Ruth Hubbard and Elijah Wald, *Exploding the Gene Myth* (Boston: Beacon, 1993), 5, 6, 8. Similarly, John Maddox contends that the "Strong Genetic Principle, that every aspect of the human condition is predetermined by the genes . . . is a fallacy." "Has Nature Overwhelmed Nurture?" 107.

52. The suggestion that XYY men are more prone to violent behavior than XY men has not been scientifically confirmed. Other more recent studies suggest possible predispositions to aggression, but confirmation is awaited. However, "accumulating evidence suggests that there is a genetic predisposition to abuse drugs [including alcohol]." John C. Crabbe, John Belknap, and Karl J. Buck, "Genetic Animal Models of Alcohol and Drug Abuse," *Science* 264, no. 5166 (June 17, 1994): 1715.

53. In this case, Baker was an attorney who misappropriated funds of his clients and was found guilty on all counts. Baker sought to avoid disbarment, and the court was sympathetic, saying that his genetic disposition to alcoholism should count in his favor. Elsewhere, exculpation has occurred. In 1994, Glenda Sue Caldwell was acquitted of charges that she murdered her nineteen-year-old son. She shot him during an angry outburst. When acquitted by Atlanta Superior Court Judge Kenneth Kilpatrick because she has the Huntington's gene, she told the Associated Press (September 29, 1994), "I was not responsible for what I did. I am a good person."

54. Maureen P. Coffey, "The Genetic Defense: Explanation or Excuse?" *William and Mary Law Review* 35, no. 353 (1993): 395–96, italics in original.

55. Ibid., 398. See also Rochelle Cooper Dreyfuss and Dorothy Nelkin, "The Jurisprudence of Genetics," *Vanderbilt Law Review* 45, no. 2 (March 1992): 313–48.

56. Charles Murray and Richard Herrnstein, *The Bell Curve: Intelligence and Class Structure in American Life* (New York: Free Press, 1994).

57. Troy Duster, "Genetics, Race, and Crime: Recurring Seduction to a False Precision," in *DNA on Trial: Genetic Information and Criminal Justice* (Plainview, N.Y.: Cold Spring Harbor Press, 1992), 132.

58. Dean H. Hamer, Stella Hu, Victoria L. Magnuson, Nan Hu, and Angela M. L. Pattatucci, "A Linkage Between DNA Markers on the X Chromosome and Male Sexual Orientation," *Science* 261, no. 5119 (July 16, 1993): 321–27. See also Simon LeVay and Dean H. Hamer, "Evidence for a Biological Influence in Male Homosexuality," *Scientific American*, May 1994, 44–45; and Dean Hamer and Peter Copeland, *The Science of Desire* (New York: Simon & Schuster, 1994).

59. William Henry III based on reports by Ellen Germain and Alice Park, "Born Gay?" *Time*, July 26, 1993, 36–39.

60. The question of whether homosexuality is or is not sinful is a matter of dispute. We are not able to settle the dispute here. Rather, we are mapping differ-

ent directions the ethical logic might take us. Pertaining to one side of the dispute, however, we note that Christian ethics has traditionally relied upon the Levitical Holiness Code, which prescribes the death penalty for homosexual acts (Lev. 18:22; 20:13); passages in the Christian Scriptures denounce homosexual relationships as idolatrous (Rom. 1:26-27) and indicate that certain kinds of same-sex practices preclude entry into the dominion of God (1 Cor. 6:9-10; 1 Tim. 1:9-10). We also note how today the Federation of Parents and Friends of Lesbians and Gays (FLAG) answers "no" to the question of its pamphlet title *Is Homosexuality a Sin?* (published in 1992 by P-FLAG, P.O. Box 27605, Washington, D.C. 20038-7605). The pamphlet quotes former Lutheran Bishop Stanley E. Olson saying, "Diversity is beautiful in creation. How we live our lives in either affirming or destructive ways is God's concern, but being either homosexually oriented or heterosexually oriented is neither a divine plus or minus." Ibid., 9. Ethicist Karen Lebacqz associates sin not with homosexuality but with "homophobia, gay-bashing, discriminatory legislation toward lesbians and gays, refusal to include lesbian/gay bisexual people into our churches and communities." Ibid., 11. James B. Nelson identifies four distinct positions within Christian ethics regarding homosexuality: (1) the rejecting-punitive position; (2) the rejecting-nonpunitive position; (3) the qualified acceptance position; and (4) the full acceptance position. "Homosexuality," in *The Westminster Dictionary of Christian Ethics*, 2d ed., ed. James F. Childress and John Macquarrie (Louisville: Westminster/John Knox Press, 1986), 273.

61. Richard Dawkins, *The Selfish Gene*, rev. ed. (New York: Oxford University Press, 1989). Speaking of evolutionary theory in light of the principle of maximum reproduction—that is, the selfish gene—William Irons notes how this fits well "with the Christian belief in original sin . . . part of the power of the belief in original sin stems from the fact that it portrays human beings as we actually experience them, that is, as having a potential for moral behavior combined with corruptibility." "How Did Morality Evolve?" *Zygon* 26, no. 1 (March 1991): 52.

62. Edward O. Wilson, *Sociobiology: The New Synthesis* (Cambridge: Harvard University Press, 1975), 4, and *On Human Nature* (New York: Bantam, 1978), 175, respectively.

63. Dorothy Nelkin and Laurence Tancredi, *Dangerous Diagnostics: The Social Power of Biological Information* (New York: Harper Collins, Basic Books, 1989), 12.

64. On the teaching of the Augustinian tradition that sin is universal though not necessary, see Reinhold Niebuhr, *The Nature and Destiny of Man*, 2 vols. (New York: Charles Scribner's Sons, 1941), 1:242. Robert John Russell helpfully delineates the Manichaean, Pelagian, and Augustinian systems. He has introduced the term "universal contingent" to describe sin as inevitable but not necessary. See his "Thermodynamics of Natural Evil," *CTNS Bulletin* 10, no. 2 (spring 1990): 20–35.

65. Augustine, "On the Merits and Remission of Sins" (*De Peccatorum Meritis et Remissione, et de Baptismo Parvulorum*), 1:16. The key to Augustine's interpretation of Paul is the unity the whole human race has both in Adam and in Christ. Once this unity is affirmed, the question becomes one of explaining it. Biological inheritance seems to serve this purpose. Yet Paul did not go that far. "He developed no theory of a biologically inherited original sin." Brevard S. Childs, *Biblical Theology of the Old and New Testaments* (Minneapolis: Fortress Press, 1993), 583.

66. Reprinted in Appendix A of Peters, *Playing God? Genetic Determinism and Human Freedom*, 179–81.

67. Gregory Fowler, Eric T. Juengst, and Burke K. Zimmerman identify three types of germ-line intervention: (1) preimplantation screening and selection of early stage embryos, which is not technically genetic engineering; (2) directly injecting normal genes into the DNA of preembryos conceived in vitro; and (3) gametocyte therapy—that is, genetic modification of gametes prior to conception. "Germ-Line Gene Therapy and the Clinical Ethos of Medical Genetics," *Theoretical Medicine* 10 (1989): 151–65. See also Burke K. Zimmerman, "Human Germ-Line Therapy: The Case for Its Development and Use," *Journal of Medicine and Philosophy* 16, no. 6 (December 1991): 594–95.

68. Christopher Wills speculates on the physical enhancements and spiritual limits of HGI: "There is nothing in principle to stop us from eventually tracking down genes that influence intelligence, skills, mental health, even behavior. But there are no genes, on any of our twenty-three chromosomes, for the soul. . . . The mere fact that we cannot prove the existence of a soul, or assign it a chromosomal location, does not invalidate that world view." *Exons, Introns, and Talking Genes: The Science Behind the Human Genome Project* (New York: Harper Collins, Basic Books, 1991), 314–15.

69. W. French Anderson, "Genetics and Human Malleability," *Hastings Center Report* 20, no. 1 (January-February 1990): 23. W. French Anderson, in a more recent work with John C. Fletcher, argues that the situation is changing. Whereas in the 1970s and 1980s there was a strong taboo against germ-line modification, in the 1990s that taboo is lifting. "Searches for cure and prevention of genetic disorders by germ-line therapy arise from principles of beneficence and non-maleficence, which create imperatives to relieve and prevent basic causes of human suffering. It follows from this ethical imperative that society ought not to draw a moral line between intentional germ-line therapy and somatic cell therapy." "Germ-Line Gene Therapy: A New Stage of Debate," *Law, Medicine, and Health Care* 20, nos. 1/2 (spring-summer 1992): 31.

70. *Manipulating Life: Ethical Issues in Genetic Engineering* (Geneva: World Council of Churches, 1981), 6–7. The 1989 document reiterates this position more strongly by proposing "a ban on experiments involving genetic engineering of the human germline at the present time." WCC, 2. Eric T. Juengst finds that the arguments for a present ban on germ-line intervention are convincing, but

he also argues that the risks of genetic accidents—even multigenerational ones—can be overcome with new knowledge. Germ-line alteration ought not be proscribed simply on the grounds that enhancement engineering might magnify current social inequalities. He writes, "The social risks of enhancement engineering, like its clinical risks, still only provide contingent barriers to the technique. In a society structured to allow the realization of our moral commitment to social equality in the face of biological diversity—that is, for a society in which there was both open access to this technology and no particular social advantage to its use—these problems would show themselves to be the side issues they really are." "The NIH 'Points to Consider' and the Limits of Human Gene Therapy," *Human Gene Therapy* 1 (1990): 431.

71. Methodist, 121.

72. UCC, 3.

73. Catholic Health Association, *Human Genetics*, 19.

74. Peter Meyer, "Biotechnology: History Shapes German Opinion," *Forum for Applied Research and Public Policy* 6, no. 4 (winter 1991): 92–97.

75. See Troy Duster, *Backdoor to Eugenics* (New York and London: Routledge, 1990), and Rifkin, *Algeny*, esp. 230–34. Not all are opposed to eugenics, especially if eugenics means better human health. For example, John Harris writes, "Where gene therapy will effect improvements to human beings or to human nature that provide protections from harm or the protection of life itself in the form of increases in life expectancy . . . then call it what you will, eugenics or not, we ought to favor it." "Is Gene Therapy a Form of Eugenics?" *Bioethics* 7, nos. 2/3 (April 1993): 184.

76. See J. Robert Nelson, *On the New Frontiers of Genetics and Religion* (Grand Rapids, Mich.: Eerdmans, 1994).

77. Methodist, 122. "Human gene therapies that produce changes that cannot be passed to offspring (somatic therapy) should be limited to the alleviation of suffering caused by disease. Genetic therapies for eugenic choices or that produce wasted embryos are deplored." *The Book of Discipline of the United Methodist Church*, 1992 (Nashville: United Methodist Publishing House, 1992), 97.

78. NCC, 21, 34.

79. Eric T. Juengst, "Germ-Line Gene Therapy: Back to Basics," *Journal of Medicine and Philosophy* 16, no. 6 (December 1991): 587–92. See also Maurice A. M. DeWachter, "Ethical Aspects of Human Germ-Line Therapy," *Bioethics* 7, nos. 2/3 (April 1993): 166–77. Nelson A. Wivel and LeRoy Walters list four arguments against germ-line modification: (1) it is an expensive intervention that would affect relatively few patients; (2) alternative strategies for avoiding genetic disease exist, namely, somatic cell therapy; (3) the risks of multigenerational genetic mistakes will never be eliminated, and these mistakes would be irreversible; and (4) germ-line modification for therapy puts us on a slippery slope leading inevitably to enhancement. They also list four ar-

guments favoring germ-line modification: (1) health professionals have a moral obligation to use the best available methods in preventing or treating genetic disorders, and this may include germ-line alterations; (2) the principle of respect for parental autonomy should permit parents to use this technology to increase the likelihood of having a healthy child; (3) it is more efficient than the repeated use of somatic cell therapy over successive generations; and (4) the prevailing ethic of science and medicine operates on the assumption that knowledge has intrinsic value, and this means that promising areas of research should be pursued. "Germ-Line Gene Modification and Disease Prevention: Some Medical and Ethical Perspectives," *Science* 262, no. 5133 (October 22, 1993): 533–38. See also chapter 6 of Roger L. Shinn, *The New Genetics* (Wakefield, R.I., and London: Moyer Bell, 1996).

80. Pope John Paul II places anthropology within creation when speaking of the humans among us as "products, knowers and stewards of creation." Robert John Russell et al., eds., *John Paul II on Science and Religion* (Vatican City: Vatican Observatory Publications; Notre Dame, Ind.: University of Notre Dame Press, 1990), M5. With reference to women in ministry, Lynn Rhodes speaks of us human beings as "co-creating the Christian vision for the future." Lynn N. Rhodes, *Co-Creating: A Feminist Vision of Ministry* (Philadelphia: Westminster, 1987), 40.

81. Paul Ramsey writes, "Men ought not to play God before they learn to be men, and after they have learned to be men they will not play God." Paul Ramsey, *Fabricated Man: The Ethics of Genetic Control* (New Haven, Conn.: Yale University Press, 1970), 138. Israeli philosopher David Heyd sees playing God more positively in terms of shared creativity: "If indeed the capacity to invest the world with value *is* God's image, it elevates human beings to a unique (godly) status, which is not shared by any other creature in the world. This is playing God in a creative, 'human-specific' way." *Genethics*, 4, italics in original.

82. See Elmer-DeWitt, "The Genetic Revolution," 48.

83. See the President's Commission for the Study of Ethical Problems in Medicine and Biomedical and Behavioral Research (Morris B. Abram, Chairman, 2000 K Street, NW, Suite 555, Washington, D.C. 20006), *Splicing Life: The Social and Ethical Issues of Genetic Engineering with Human Beings* (Washington, D.C.: Government Printing Office, 1982), 54.

84. Ibid., 95–96.

85. Zimmerman, "Human Germ-Line Therapy," 606.

86. Robert L. Sinsheimer, "Genetic Engineering: Life as a Plaything," *Technology Review* 86, no. 14 (1983): 14–70.

87. Rifkin, *Algeny*, 17.

88. Ibid., 252. Although the phrase "play God" has been with us for some decades as a reference to the prospect of scientific creation or manipulation of life, Je-

remy Rifkin thrust it before the public with his 1977 book title *Who Should Play God?* (New York: Dell, 1977). Rifkin has garnered his share of critics. Walter Truett Anderson dubs Rifkin's hysterical attack against genetic engineering "biological McCarthyism." Anderson's position is that the human race should become deliberate about the future of its evolution: "This is the project of the coming era: to create a social and political order—a global one—commensurate to human power in nature. The project requires a shift from evolutionary meddling to evolutionary governance, informed by an ethic of responsibility—an evolutionary ethic, not merely an environmental ethic—and it requires appropriate ways of thinking about new issues and making decisions." *To Govern Evolution* (New York: Harcourt, Brace and Jovanovich, 1987), 9, 135.

89. Ronald Cole-Turner, *The New Genesis: Theology and the Genetic Revolution* (Louisville: Westminster/John Knox Press, 1993), 45.

90. Ibid., 45.

91. Hessel Bouma et al., *Christian Faith, Health, and Medical Practice* (Grand Rapids, Mich.: Eerdmans, 1989), 4–5. James M. Gustafson makes "the theological point that whatever we value and ought to value about life is at least relative to the respect owed to the creator, sustainer, and orderer of life." "Genetic Therapy: Ethical and Religious Reflections," *Journal of Contemporary Health Law and Policy* 8 (1992): 196. For Gustafson, the central question around which the issue of germ-line intervention is oriented is this: How do we define what is naturally normal for human life? For the theologian to answer this question, more than knowledge of biology is required. Required also is awareness of the divine ordering of human life.

92. Ted Peters, *GOD—The World's Future: Systematic Theology for a Postmodern Era* (Minneapolis: Fortress Press, 1992), chap. 4.

93. Philip Hefner, "The Evolution of the Created Co-Creator," in *Cosmos as Creation: Science and Theology in Consonance*, ed. Ted Peters (Nashville: Abingdon, 1989), 211–34; and Philip Hefner, *The Human Factor* (Minneapolis: Fortress Press, 1993), 35–42.

94. In addition to ethical responsibility, church leaders anticipate a broader pastoral responsibility. Ethicist Karen Lebacqz recommends that congregations respond to new developments in genetic therapy through (1) support for persons and families facing illness and death; (2) education of and by the clergy to help all church members interpret genetic illness and its accompanying hopes and tragedies; and (3) advocacy in public policy debate that will support legislation to aid disabled persons and others needing social protection. Karen Lebacqz, ed., *Genetics, Ethics, and Parenthood* (New York: Pilgrim Press, 1983).

95. Thomas A. Shannon, *What Are They Saying About Genetic Engineering?* (New York: Paulist Press, 1985), 94.

96. UCC, 2; WCC, 31; Methodist, 114; Church of the Brethren, 453; United Church of Canada, 14.

97. United Church of Canada, 14.

98. Ibid., 13. See NCC, 43; Lutheran, 2; *Manipulating Life*, 7–8.

99. Pope John Paul II, "Biological Experimentation," address given October 23, 1982; published in *The Pope Speaks*, 1983.

100. Cole-Turner, "Genetics and the Church," 55.

PART ONE

GENES AND JUSTICE

[2]

The Genome and the
Human Genome Project

R. David Cole

THE TASK OF THIS chapter is to introduce the basic science of genetics and to therein identify the goal of the Human Genome Project (HGP). In the first section we will trace the process by which genes express themselves and describe the key concepts with which biologists at the molecular level work. In the second section we will turn to the specific goal of the international enterprise known as the Human Genome Project, namely, to map the base sequences in human DNA and locate all the genes. The goal is to gain knowledge about human DNA. Yet the widespread public support for the HGP comes from a related goal, namely, better human health through science.

Let us start with curiosity, the stimulus for pursuing knowledge. To ponder a newborn baby, perhaps with its mother's eyes and its father's chin, is to face a profound puzzle. Why does it look so much like all other babies and yet not quite like any other in the whole world? Why does it resemble its father in some particulars and its mother in others, and in some particulars look more like an aunt or an uncle than it does like either parent? We see nearly measureless diversity in human characteristics, but there are evidence of order and a sense of highly specified structural design. The human genome is the chief source of information for such designs. Moreover, the importance of passing genomic information from one generation to the next goes far beyond mere appearances. The baby works the everyday magic of transforming the few dozen kinds of chemicals that compose milk into the tens of thousands of chemical species needed by the baby to build its body and to manipulate it. It does not learn to do that; it inherits the design and a tool kit with which to craft its marvelous machine. The power and finesse to perform such wonders as the chemical transformations are manifestations of exceedingly complicated and finely structured machinery. In order to understand the design and workings of that machinery, the HGP was undertaken by the United States and joined by many other countries.

What Is the Human Genome?

Phenotype, the Work of Proteins

The extremely specific structures and processes of the human body are largely the work of proteins, even though the design rests in the genome. In general, each structure and each process is the handiwork of not one but a number of proteins. Thus, the phenotype of any human, all observable characteristics, results from the structure and function of a vast array of proteins.

The structure of an average protein at the primary level is a string of about three hundred subunits of twenty kinds. The twenty kinds of subunits, called amino acids, are lined up in a particular sequence that differs from one protein to the next. A protein may be compared to a paragraph of prose in which letters and spaces are lined up in a particular sequence. Since the twenty kinds of amino acids differ among themselves in chemical character, every kind of protein differs in chemical character from others just as different paragraphs in a novel differ from one another. To accomplish the complicated work of the human body, tens of thousands of proteins are needed. In addition to that complexity, when proteins are synthesized at the primary level as a string of connected amino acids in a particular sequence, the string folds into a definite, three-dimensional structure with precise geometry that differs from one protein to the next. The final structure of a protein can tolerate replacement of some of its amino acid subunits with others of a different kind even if this results in a modest alteration in three-dimensional folding. The degree of tolerance for such changes (mutations) is small and differs from one part of the protein structure to another. Changes may be lethal, they may partially inactivate the protein, or in some cases they may have no substantial effect. Rarely is protein function enhanced by a mutation.

The limit to which proteins can tolerate mutations is further restricted by the fact that most proteins are designed to coordinate their activities with those of other proteins. The integration of the functions of many proteins into large networks allows the organism to shift its balance of activities according to changes in the demand for the system's output. In short, such networks allow the human to adapt to environmental change. The sensitivity of each protein to the presence or function of another protein requires that the design of both proteins specifies structures that not only perform the primary task of the protein, but respond to chemical signals as well.

The Genome Is DNA

The design for the highly specified structures of proteins, which must anticipate their fundamental action and their integration into a network of other proteins, is carried in deoxyribonucleic acid (DNA). Like proteins, DNA molecules are strings of subunits of different kinds arranged in particular sequences that spell out the instructions for making various proteins. In the case of DNA, there are only four kinds of subunits, and they are called nucleotides. All four kinds of nucleotides have three connected parts: a sugar, a phosphate residue, and one of four kinds of bases—adenine (A); thymine (T); cytosine (C); or guanine (G). As will be shown shortly, a triplet of nucleotides of a particular sequence, called a codon, can be translated into one kind of amino acid. Thus, the message for the design of a protein is spelled out in DNA as a series of nucleotide triplets and subsequently translated into the series of amino acids appropriate to the protein being synthesized.

To string the nucleotides together, the phosphates form bridges between the sugars of the nucleotides, leaving the bases to dangle from the chain of alternating phosphates and sugars. The DNA molecule has two strings of subunits. The two strands of nucleotides in DNA are entwined around each other, each with its strong chain of alternating sugars and phosphates on the outside of a helix, and with the bases pointed to the inner axis. Each base on one strand of the DNA, strongly bound to a sugar in its own strand, is more weakly bound to a base on the other strand. Because of the shapes and chemical character of the bases, the pairs of bases facing each other are always either A and T, or C and G; no other combinations fit together. The two strands of DNA are complementary to each other, and the sequence of bases in one strand can be deduced from the sequence in the other strand. This complementary base pairing of nucleotide strands provides the basis for passing messages from one generation of organisms to the next and for translating messages from nucleotide language into amino acid language in the synthesis of proteins. The inside of the double helix formed by the two intertwined nucleotide strands could be likened to a spiral staircase where each step is a pair of complementary bases. The number of base pairs (bp) in molecules of human DNA varies from 50 million to 250 million; and there may be up to 3 billion base pairs comprising the totality of human DNA.

Transmitting Genomic Designs from Cell to Cell—Replication

During the growth of a human, and during adult life when worn cells need replacement, new cells must be made, and this is accomplished by division of a parental cell into two progeny cells. Since each progeny cell receives a full set of instructions for making the constituents of the whole body, the DNA of the parent cell must be replicated. Shortly before cell division, the two strands of DNA are unwound, and the strands are separated. While separating, each strand becomes a template for the synthesis of a new complementary strand by lining up individual, complementary mononucleotides and stitching them together by a protein called DNA polymerase to form a polynucleotide strand. The final result is two double helices, each identical to the original, one for each of the progeny. Each progeny cell gets one parental polynucleotide strand and a complementary new strand.

Transmitting Information from Genome to Proteins—
Transcription and Translation

The human genome is the complete set of 50,000 to 100,000 human genes, each of which is a region of DNA that contains the information for the controlled biosynthesis of a gene product—usually a protein, but in some cases an RNA. Along with DNA used in regulation of the biosynthesis, each gene possesses a coding region that ultimately is translated into the structure of one protein (or RNA). The synthesis of a protein, according to the instructions encoded in its gene, involves an intermediate level of information transfer. Because the DNA is confined to the nucleus of the cell, while the protein-synthesizing machinery—the ribosome—is located outside the nucleus, a messenger is used to carry the code from the nucleus to the ribosome. This messenger is called messenger RNA (mRNA). The RNA is a polymer of nucleotides much like DNA except that its sugar residue is slightly modified, and it is single stranded rather than double stranded. Like DNA, RNA has four bases, but it uses uracil (U) in place of the thymine of DNA. It speaks the same language as DNA in that it can form complementary base pairs with DNA with the pairing of C and G, A and U, and T and A. Therefore, when genomic information is passed from DNA to RNA, it is called transcription, whereas transferring information from the nucleotide language of RNA to the amino acid language of proteins is termed translation. In transcription, one strand of the DNA serves as a template to line up a sequence of complementary nucleotides, which are then stitched together into a long strand of mRNA by a protein called RNA polymerase. The mRNA is released from the DNA, exits the nucleus, and eventually binds to a ribosome. The ribosome picks up appropriate amino acids from the surrounding medium, stitching one amino acid to the next as it works its way along the length of the mRNA, and translating the sequence of nucleotide triplets into a corresponding sequence of amino acids.

Regulation of the Flow of Information

The cell's DNA contains more than the nucleotide sequences that code for amino acid sequences. For one thing, a (small) fraction of the DNA encodes RNA structures that are not translated to proteins. A major portion of the DNA does not have a recognized function and is frequently referred to as "junk DNA," but the term "junk" may not be apt. Some of this DNA may play a structural role, and some may serve as a reservoir of nucleotide sequences useful in long-term responses to environmental changes in the evolutionary sense.

An especially interesting kind of noncoding DNA is used in the control of gene expression. The cells of a highly developed organism such as the adult human are so specialized in function that they need to implement only a subset of the genes in the genome. Most of the genes need to be repressed. Of the genes that are expressed, the extent of gene expression will need to be modulated to control protein production as a person adapts to circumstances such as a change in altitude, a regimen of vigorous exercise, or a fat-rich diet. Each gene therefore carries "regulatory" sequences designed to interact with regulatory molecules, usually proteins, in ways that turn gene expression on or off, or that adjust the rate of transcription into mRNA. The regulation of gene expression in humans is by no means well understood, and the determination of the structure of the regulatory sequences in genes is an important part of the HGP.

Transmitting Genomic Designs from a Parental Cell to Progeny Cells

During most of its life, the human cell has its genome in a loose complex of protein and DNA, but when a cell divides to form two progeny cells, it must replicate its genome and package it into bundles that can be neatly dispatched into one progeny cell or the other. The human genome is organized into forty-six particles called chromosomes, each of which is a single piece of DNA. There are actually twenty-two pairs of homologous chromosomes (autosomes) that account for forty-four of the particles. The other two particles are a homologous pair of X chromosomes in the female, or a heterologous pair in the male, an X chromosome and a Y chromosome. One member of each pair was derived from the father of the person and the other member from the mother, although the two members of a homologous pair are very much alike but not identical because the genetic heritage of the father was not like that of the mother. The Y chromosome is substantially smaller than the X chromosome and contains genes that are homologous with some of those in the X chromosome. Each homologous chromosome may have a slightly different version of any particular gene. Each version, or variant, is referred to as an allele of the gene. If the two alleles of a gene are identical in the homologous pair, then the person is termed homozygous, but if the alleles differ, the person is heterozygous.

In the case of an autosomal pair, an allele with normal function in one of the chromosomes can frequently compensate for a deficient allele in the other, but for many genes in the heterologous pair of chromosomes in males, compensation is impossible. Y does not carry genes corresponding to those in X.

Just before cell division, all the DNA in the cell is replicated (see above) making forty-six pairs of chromosomes, duplicate sets of twenty-three pairs. Each chromosome condenses adjacent to its duplicate but not usually near its homologous mate. A special apparatus is constructed in the cell at the appropriate time to pull one set of twenty-three pairs to one end of the cell and the duplicate set to the other end. When the parent cell divides, each of the progeny cells possesses a set of chromosomes identical to the parental set.

Transmitting Genomic Designs from Human Parents to Children

Passing genomic designs by sexual reproduction is more complicated than the process just described for simple cell division. Mendel's laws of inheritance reveal that a child receives one chromosome of every pair from each parent; therefore, the specialized cell that is the sperm will carry twenty-three unpaired chromosomes, as will the egg cell. When sperm meets egg, it injects its chromosome into the egg, which thus fertilized will contain twenty-three pairs of chromosomes. Sperms and eggs with sets of unpaired chromosomes are produced in a special process called meiosis. During meiosis, sections of one chromosome may exchange with homologous sections of another chromosome. The resulting sperm or egg chromosomes contain alleles that have crossed over from a homologous chromosome. As would be expected, genes close to each other on a chromosome will cross over together more often than distant genes will. Measuring the distance between genes by the frequency with which they cross over together is the basis of genetic mapping.

Goals and Strategies of the Human Genome Project

The understanding of molecular genetics sketched in the previous section was accumulated over the last century—and especially since 1953—when the double-stranded, complementary based–paired structure of DNA was recognized. In the 1970s and 1980s, the technology was developed for reading the sequence of bases in DNA and for general genetic engineering. Taken together, these advances revealed the fundamental importance and technical feasibility of learning the structure of the genome in complete detail. In 1988 the United States, soon joined by other countries, undertook the Human Genome Project. In its initial articulation, the goals were to learn the sequence of all three billion bases in the DNA of the human genome and to locate reference points spaced about 100,000 bp apart all along the sequence (i.e., to map the genome). Although not specified in the original goals, it was anticipated that analysis of the data obtained in the HGP would continue through the twenty-first century, leading to new strategies of diagnosis and treatment of the few thousand known simple genetic diseases, and perhaps many others in which genes are contributory factors.[1]

Genetic Mapping

The position in the chromosome is known for only a minor fraction of the genes that have been identified by protein product or by inherited trait; two to three thousand genes have been located approximately, usually within a million or so nucleotide bases. The base sequence has been elucidated for only a fraction of the located genes, but both sequence and position data are rapidly accumulating for DNA of currently unidentified function.

The search for the chromosomal position of genes began in the 1910s when the concept of genes as paired units of inheritance was associated with accumulating knowledge of paired chromosomes seen microscopically in dividing cells. At this stage in the life of a cell, chromosomes are condensed into woolly cylinders of various lengths, and if stained with dyes, each chromosome shows a characteristic pattern of light and dark bands in the microscope. The members of a homologous pair of chromosomes have the same length and banding pattern, but are unlike other chromosomes. Genetic mapping became a matter of discerning which band contained the gene of interest.

Some inheritance statistics of the eye color of fruit flies were found not to show a pattern of inheritance in succeeding generations like the pattern for seed color described by Mendel. Mendel's patterns were based on genes that occur in homologous pairs so that each generation inherits the paternal genes and the maternal ones in equal numbers. The numbers were not equal for the inheritance of red eyes in fruit flies but could be explained if the gene for eye pigment was absent from the Y chromosome that was unique to males. Therefore, the gene for eye pigment must be located in the X chromosome. Following this first step in the location of a sex-linked characteristic, several other genes were located on the X chromosome.

As noted previously, the process of meiotic crossover provides a way to measure the distance between two genes. During meiosis, a section of one chromosome—carrying one allele of every gene in that section—may exchange with the corresponding section of the homologous chromosome, which contains its alleles of those genes. If two genes are close to each other, their alleles are more likely to cross over together than if they are far apart, and so the frequency with which the two alleles are found to be separated in patterns of inheritance corresponds to the distance between them. The closer two genes are, the more frequently their alleles will be coinherited. Genes separated by crossing over 1 percent of the time would be roughly one million base pairs apart. For more than two thousand cases in which a single gene is responsible for a disease, inheritance patterns have allowed the placement of the responsible genes as reference points (markers) in chromosomes. Other markers can be located by inheritance patterns when minor differences in base sequence (polymorphisms) can be easily detected between paternal and maternal alleles. Mapping by inheritance patterns permits markers to be located only within about ten million base pairs.

What Is the Human Genome Project?

Physical Mapping

One goal of the HGP is to make chromosome maps with markers spaced about every 100 Kbp apart, so that locating a gene between two adjacent markers will fix its position quite closely.[2] Although some markers are being placed by crossover studies, most are being located by physical mapping of chromosomes in which physical and chemical measurements are made between markers. Distances are thus measured directly as base pairs or as an actual distance observed in the microscope.

The microscope can be used to observe markers with a resolution of a few million base pairs (Mbp). A single stranded DNA fragment can be tagged with a fluorescent label in order to be seen under the microscope. If an array of chromosomes is exposed to the fluorescent DNA fragment, the fragment will bind to the chromosome where it finds a complementary nucleotide strand. The site of binding marks the chromosomal position of that sequence. A common type of DNA to use as such a probe is cDNA. cDNA is produced by treating an mRNA with the enzyme reverse transcriptase to make a single strand of DNA with a base sequence complementary to that of the mRNA. Since the mRNA is complementary to one of the strands of its gene, the binding site for the cDNA marks the chromosomal site of the gene. Unfortunately, the precision of this method is limited by the resolving power of the microscope, so that the positions of genes or other markers lying closer than a few Mbp cannot be distinguished from one another.

Higher resolution can be obtained in physical mapping by fragmenting DNA with two or more restriction nucleases. A restriction nuclease breaks DNA strands at places containing a particular multibase sequence. The cleavage sites recognized by restriction nucleases are usually four to eight nucleotides long, but a few recognition sites are even longer. The longer the recognition site, the more rarely that particular base sequence will be found in DNA, and therefore, the fewer will be the fragments when DNA is cut by that restriction nuclease. The strategy for mapping genomic DNA with restriction nucleases is to make a first series of fragments by cutting the DNA at one set of restriction sites, and then to compare that series to a second series made by fragmenting the original DNA with other restriction nucleases. If markers (e.g., cDNA binding sites) in the tail of a fragment in the first set overlap markers in the head of a fragment of the second set, the two fragments must be derived from tandem positions in the original DNA. When many fragments can be overlapped to reveal their positions in a long stretch (called a contig) of original DNA, the markers in the fragments are arranged as a physical map. Since the base sequence of three hundred to five hundred nucleotides at one end of each fragment is easily determined, detailed structural information is at hand for every overlap region, and that sequence becomes a new marker.

Mapping with restriction fragments is done in two modes. In one, a relatively few large fragments are prepared using a restriction nuclease specific for a rare (i.e., long) nucleotide sequence. The second mode is to shatter the DNA into many small fragments with nucleases that recognize more common (i.e., short) base sequences. The first mode has advantages complementary to those of the second mode. In the first mode, the small number of fragments simplifies the purification of individual fragments for further study (e.g., identification of markers), and the small number also profoundly simplifies the process of finding overlaps. The disadvantage is that the small number of overlaps leaves markers widely spaced and yields only a small number of regions where the three hundred to five hundred bp sequences can be analyzed.

The second mode of mapping with restriction nucleases produces many small fragments and, therefore, many regions where base sequencing can be done. However, the large number of fragments makes it difficult to overlap a series of fragments into long contigs. The larger number of fragments also makes it extremely difficult to purify sufficient amounts of individual fragments by physic-chemical methods. Cloning, a biological method, has therefore been used for many years to purify practical quantities of fragments. But before I describe cloning, I ought to mention a chemical process that can amplify tiny amounts of particular DNA fragments—the polymerase chain reaction (PCR).

To synthesize a new strand of DNA, the enzyme DNA polymerase needs a primer oligonucleotide of ten bases or more that are complementary to the template DNA where the synthesis is to start. This oligonucleotide can be chemically synthesized, or purified from natural sources, and the fragments used in gene mapping can serve as primers. The PCR3 is a cyclic process that doubles the template DNA at each cycle, amplifying the original amount of DNA to any desired level by extending the number of cycles. The impact of PCR on the rate of progress of the HGP has been enormous in preparing practical quantities of DNA, but even more in its analytical use, where it can amplify an unknown chromosomal DNA that is complementary to a cDNA of interest or to a synthetic oligonucleotide "probe" if either is used as a primer.[3]

Cloning DNA Fragments

The strategy of purification by cloning is based on the idea that if a population of millions of bacteria incorporates thousands of DNA fragments and replicates them, most of the bacteria will incorporate and replicate only a single fragment. If the bacteria are widely scattered onto a nutrient-containing gel, each bacterium will proliferate by repeated cell divisions, becoming a colony (clone) of identical progeny. Each progeny cell will have a copy of DNA identical to that of the original bacterium, including any DNA fragment that was incorporated into the host bacterium. The DNA fragments that were mixed together at the beginning are now all separated from each other, each in a different clone of bacteria. Each clone can be picked off the surface of the gel, placed in a nutrient-containing medium, and allowed to multiply by itself. In this way, single kinds of DNA can be grown to quantities adequate for chemical analysis. Large collections of clones representing, for example, chromosome 21, are called libraries.

To incorporate DNA fragments into bacteria, they are transported by vectors that are either plasmids or viruses. Plasmids are circles of double-stranded DNA that can penetrate bacterial cell walls. There are naturally occurring plasmids, and artificial ones can be tailor-made for the purposes of genetic engineering. Viruses are particles in which a piece of nucleic acid is surrounded by a protective coat of protein. They are naturally designed so that the protein binds to a cell wall and injects the nucleic acid into the bacterium. Plasmid or viral DNA can be used as a vector for transporting DNA fragments into bacteria by another application of restriction nucleases. The plasmid or viral DNA can be broken open with the restriction nuclease, and a fragment of foreign DNA is stitched into the opening with an appropriate enzyme. That product can then be taken up by a bacterium. DNA fragments up to about 15 Kbp can be inserted into plasmids this way. Similarly, larger fragments—up to about 40 Kbp—can be inserted into an artificial vector constructed from the cos gene of a virus, and hence called a cosmid. Still larger fragments—up to about 1000 Kbp—can be inserted into a yeast artificial chromosome, the so-called YAC, constructed from critical parts of the yeast chromosome.[4] Variations of these techniques also allow the cloning of DNA fragments into the cells of higher organisms (e.g., humans) when the cells are cultured in vitro. Bacteria, however, are much easier to grow in large quantity; therefore, they are used to prepare enough of a DNA fragment to permit the determination of its base sequence.

The introduction of YAC technology had a major impact on the speed of mapping because it allowed much larger pieces of DNA to be studied, thus making the overlapping of fragments much simpler. Combining the use of YACS, with PCR in the analytical mode, along with the development of robotics on a large scale enabled Ilya Chumakov et al. in France to complete a physical map[5] of the transcriptionally expressed part of chromosome 21 by October 1992. At the same time in the United States, Simon Foote et al. used a similar approach to complete a similar map[6] for the expressed part of the Y chromosome. Both maps showed markers about 200 bp apart. These achievements vindicated what had been a controversial approach. They worked out so much faster than expected that the National Institutes of Health (NIH) decided to develop a factory-scale center[7] like Daniel Cohen's (Chumakov et al.) in France—a major expansion of Eric Lander's relatively modest laboratory. The work in Cohen's center also convinced him that the genome could be mapped as a whole. Previously, chromosomes were studied individually to simplify the analysis; sorting chromosomes is a slow and laborious process. The validity of Cohen's view was demonstrated in October 1993 when he announced a nearly complete map[8] with marker spacings averaging 1 Mbp. Although it is still low resolution, it puts the HGP well ahead of schedule, and in adding new markers, many more researchers will work with the whole genome.

Incidentally, this story is one example among others of a magnificent spirit of cooperation that has developed throughout much of the community of genome researchers. Not only has Cohen made his data available to others quickly, but he has made his library of thirty thousand cloned YACS available to others.

Base Sequence Determination

Currently, there are two methods for determining nucleotide sequences. One approach is to synthesize new complementary strands beginning at one end of a single strand of the DNA whose sequence is to be determined, and to stop the synthesis in a small fraction of the new strands every time the growing strand comes to a particular kind of nucleotide—for example, at each C. Ultimately, a mixture of new strands is formed in which the strands differ in length; each length corresponds to the distance from the start of the synthesis to one of the C's in the DNA whose base sequence is being sought. Since there is a technique (electrophoresis) that measures the length of all the fragments in a single experiment, the position of every C can be determined easily.

The other approach to sequencing bases uses chemical cleavage specific for one base or another. For example, first the DNA strand of unknown sequence is radioactively or fluorescently labeled on one end, and then partially cleaved at its C bases. A mixture of fragments is produced from all the combinations of partial cleavages at all the C's. The set of fragments still bearing the radioactive or fluorescent label will form a series of fragments of increasing size corresponding to distances from the original end of the unknown DNA to each C in the overall sequence. Both methods are limited to the determination of the first three hundred to five hundred bases of any fragment by the nature of the method used to measure fragment lengths. This limitation must be overcome or circumvented if the HGP is to reach its goal on time.[9]

Diagnosis of Genetic Diseases

Once a gene is located and isolated by cloning, its structure can be compared from one person to another, and differences may be observed in nucleotide sequence. Such mutations sometimes have little or no effect on the gene product (protein), but in many cases a mutation will result in a nonfunctioning or malfunctioning gene product, and in general will cause or contribute to a disease. Although not an explicit goal in the original research plan of the HGP, the hope was expressed that the knowledge of gene structure would eventually aid the diagnosis of genetic diseases. Thus, genetic counseling practice would be greatly benefited. Certain genetic diseases can be circumvented or ameliorated by early treatment, and in such cases early diagnosis would be especially valuable.

Identification of genes is going forward at least as fast as had been hoped initially, and as scientists anticipated, this knowledge is enriching and stimulating biological research beyond the immediate goals of the HGP. A few discoveries that caught widespread public attention are genes related to Huntington's disease,[10] Lou Gehrig's disease,[11] childhood leukemia,[12] Alzheimer's disease,[13] fragile X mental retardation,[14] Duchenne muscular dystrophy,[15] and cystic fibrosis.[16] In some of these cases, identification of the gene has led to a fuller understanding of factors and processes involved in normal as well as diseased states.[17] The identification of genes responsible for Huntington's, fragile X, myotonic dystrophy, and bulbar muscular atrophy revealed a totally unforeseen phenomenon—tandem, repeated nucleotide triplets varying in number of repeats. How the repeat number is varied and how it affects phenotype present a new puzzle for basic researchers apart from those directly related to HGP work.[18] These advances are both exciting and rewarding, but they also have made clear that application of human genome knowledge will be exceedingly complicated.

Diagnoses might be relatively straightforward in diseases caused by the malfunction of single genes, but relatively few genetic diseases are monogenic. Most genetic diseases involve many genes, and an understanding of them would require simultaneous consideration of all the genes and nongenetic factors participating in the disease phenotype.[19] If only one of a set of participating genes is recognized as correlating with the disease, and others left out of consideration, poor advice might be given to patients. Even in the case of a monogenic disease, complications can arise in screening people for the mutant genes. One of the triumphs of gene hunters was the location of the gene related to cystic fibrosis. Although a single gene and a single gene product are mutated in this disease, the mutation occurs in more than 360 ways—360 places in the gene where the base sequence is altered.[20] A screening probe specific enough to recognize one particular change in base sequence will fail to detect all the other mutations, and a probe general enough to include all the mutants will fail to distinguish between the mutants from normal genes. Fortunately in this case, it might prove possible to group the mutants into a half dozen major classes for which class-specific probes could be developed. Another monogenic disease that has given trouble is hemophilia. Most people with hemophilia have a mutated gene for Factor VIII, an enzyme of the blood-clotting system. In screening these people by determining whether their DNA would show complementary base-pairing with normal DNA, the DNA of about half of them paired with normal DNA, indicating that they contained the same base sequence as in the normal gene. What was eventually discovered[21] was that in the DNA of the people with hemophilia, the normal sequences were indeed present, but running in the reverse direction from normal, confounding proper transcription.

The situation is further complicated by the fact that many—perhaps most—gene products participate in large networks, the function of which must be regulated in order to adapt to changes in circumstances. The components of such networks are commonly affected by multiple, widespread factors through elaborately indirect routes. This might well apply to the case of inherited susceptibility to cancer. An example is colon cancer, where nine genes have been identified that show a significant degree of correlation with susceptibility to this cancer.[22] Functions of identified genes related to cancers are quite varied: cell adhesion, signal reception in membranes, phosphorylation of enzymes, transcription regulation, and so on.[23] A similar prospect is faced in genes associated with a heritable form of Alzheimer's disease, where such genes have been identified on four different chromosomes.[24] The opposite side of the coin is that a single component in a network of feedback loops can be involved in numerous pathological states. This might turn out to be the case for the monoamine oxidase gene, the loss of which has recently been claimed to be related to violent behavior. This enzyme functions in the metabolic inactivation of neurotransmitters and related metabolic problems, and has also been implicated in attention deficit hyperactive disorder, Tourette's syndrome, and alcoholism.[25] Not only the large number of biological components (gene products) must be considered in trying to understand a disease condition, but account must be taken of the dynamic balance of these networks and the effects of nongenetic and environmental factors on that balance. What fraction of genetic diseases can ultimately be understood largely in terms of mutations in gene structure remains to be seen. It is intimidating to face the emotional and psychic trauma that will probably develop in the public while researchers stagger to an understanding of genes that are claimed to be related to such touchy issues as homosexuality and violence.[26]

Although many people may be disappointed to find many diseases still uninterpretable when the structure of the human genome is fully known, at least the thousands of patients suffering from the simpler, understandable genetic diseases will be well pleased with the efforts of the HGP. Even if the number of diseases that yield their secrets by means of genetic structure might be small on a percentage basis, the absolute number may be substantial enough.

Gene Therapy

The dream of supplementing, or even replacing, faulty human genes came into public discussion twenty years ago when we first learned how to redesign genes in the test tube with restriction nucleases. The general approach to gene therapy was evident even then, and most of the experiments on gene transfer to date have followed it.[27] Differentiated cells from the target tissue of the patient are cultured in vitro, and the gene of choice is incorporated (transduced) into them. The transduced cells are then injected back into the patient where they supplement faulty gene product with their correct one. There are three limitations to this approach. The first is that differentiated cells do not reproduce themselves, and so their effect is transient. The second limitation is that the injected cells may not be located and retained where their gene product is needed. The third limitation is inherent in the high degree of specialization of fully differentiated cells, whose genome is about 85 percent regulated in the "off" mode. The engineered gene must express itself in the face of the cell's regulatory system as that system responds to the environment it encounters after injection into the patient.

One way to overcome the limitations of using differentiated cells for gene therapy would be to incorporate engineered genes into ova. Once fertilized, an ovum develops to become the whole body with all its cell types. By the very nature of the developmental process, all the regulatory systems and feedback loops develop for normal function of the gene in its normal locations. Furthermore, as cells wear out they would be replaced in the normal way with cells containing the engineered genes. Of course, this mode of therapy applies to future generations rather than an existing one.

Although contrary voices are increasingly heard,[28] there has been a consensus against engineering genes into the germ line because it would commit unconsenting future generations to a change in their heritage. Instead, the focus has been on transferring genes into the fully differentiated cells that make up most of the body, or into stem cells. Fully differentiated cells are highly specialized as nerve cells, insulin secretors, or muscle cells, among others, and they wear out on a timescale of several months to several years, to be replenished by the growth (cell division) of stem cells. Stem cells, which comprise a tiny percentage of our total cell mass, have been interrupted in their differentiation. Stem cells in the bone marrow, for example, can finish their differentiation to become any one of several types of cells that circulate in the bloodstream. They are too differentiated to become like most of the other kinds of cells in the human body; they are committed to become any one of a small set of blood-related cells. Which member of that small set is determined as the need for cell replenishment arises from the wearing out of old cells. For purposes of gene therapy, the introduction of healthy genes into a patient's stem cells has the advantage that the injected cells would persist indefinitely, producing good gene product. Introduction of the genes into a fully differentiated cell has the disadvantage that the cells wear out and disappear so that the benefits of their good gene product would be transient. Unfortunately, isolation of amounts of stem cells sufficient for the manipulations of genetic engineering is enormously difficult, while fully differentiated cells are readily available. Trials[29] with stem cells have given encouraging results, and it is to be hoped that the technology of stem cell isolation can be made more practical.

Yet another candidate for gene therapy is the fetal cell.[30] Although much more differentiated than ova, fetal cells are not fully committed to a single cell type, and in any case they can proliferate indefinitely. Experiments in model animal systems have demonstrated the effectiveness of this approach, but its use will face strong public opposition.

Once the cell type is chosen, a method is needed to transport the engineered gene into the cell. Most commonly, viruses have been used as vectors since it is in their nature to attach to the surface of cells and inject their nucleic acid into the host cell to exploit the synthesizing machinery of the host in the production of new viruses. If a gene of choice is engineered into the viral nucleic acid, it can go along for the ride into the cell, and it, too, will be transcribed and replicated. The viral vectors being used in clinical trials have been engineered to lack some component critical to viral reproduction without impairing the ability of the virus to inject its nucleic acid into the cell, but some concerns remain for slight, residual infectivity. Some researchers are trying to develop viral vectors that would self-destruct upon (chemical) command.

In several experiments, viruses have been used to transduce bone marrow cells from patients, introducing healthy genes to supplement the faulty genes. The genes engineered into the viruses were ones that normally produced blood proteins and immune proteins. When the transduced cells were injected back into the patient, the desired gene products were, in fact, synthesized. The levels of gene expression were quite low, however, and the levels fell rapidly with the cells persisting only a matter of weeks. The low level of expression revealed, among other things, that pilot studies done to estimate the level of gene expression in standard cells cultures were not reliable guides to the expression rate in vivo. Current research on hemopoietic growth factors might provide a solution to the problem of low gene activity. Research is under way on the transduction of genes into several other cell types, and on the use of nonviral vectors such as lipid vesicles that can fuse with the lipoidal cell membrane, releasing its contents into the interior of the cell.

Clearly, gene therapy is in the very early stages of development. Despite its great challenges, its potential is so attractive that many research centers have been established for relevant research, with heavy financial support from commercial concerns as well as the government.[31] At the start of the HGP, the hope was that commercial interests would take over much of the burden of developing gene therapy; no one imagined that it would happen for several decades. There has been so much excitement, however, that the investment of private concerns into gene therapy research for 1994 was estimated to equal that of NIH. This has occurred when the only success that can be claimed is the transfer of healthy genes for the enzyme adenine deaminase.[32] A deficiency of this gene is a rare genetic disease that severely impairs the development of the immune system so that its victims usually die of infection at an early age. Two young girls afflicted with this disease were given gene therapy of the sort described above; now they are leading apparently normal lives and attending public schools. It is an inspiring story! In any case, commercial interests have rushed in to develop this application of information from the HGP, and this commercialization may spare taxpayers some of their burden, although it raises questions about the ultimate distribution of benefits.

Shifting Strategies

The five-year research plan of the HGP laid out in 1990 was replaced on October 1, 1993, by a new five-year plan.[33] The revision was a response to unexpectedly rapid advances in certain aspects of research as well as to reactions of the broader scientific community to the HGP.[34] The original goals for the overall project were a genetic map at 2 Mbp resolution; a physical map at 100 Kbp resolution; a library of cloned fragments about 5 Kbp in length; complete base sequence of the genome; the finding of all the genes; and an informatics system for receiving, processing, and transmitting information. With the data generated the broader scientific community was expected to determine functions of the genes and develop medical applications.

Although the British decided to work mainly on the structure of just the expressed genes (cDNA), the American leaders committed themselves to sequencing the entire DNA. The U.S. project also emphasized a decentralized, "little science" approach—the use of many small laboratories rather than a few large ones. To accomplish the project on time and within budget, it was anticipated that other countries would join the United States, and that there would be an especially robust spirit of cooperation throughout the research community. It was further anticipated that enormous advances would be made in technology—not just improvements of current methodologies, but breakthrough innovations. Increases in the speed and cost of data acquisition, analysis, and distribution of data needed to improve by multiple orders of magnitude. At the outset opponents of the HGP felt these hopes were grossly unrealistic. Actually, by the mid-1990s the project was not behind schedule in any of its parts, though there were concerns that base sequencing and "informatics" might soon face serious difficulties.

Genetic mapping is not only ahead of schedule but under cost, and physical mapping is at least up to schedule. Progress was profoundly accelerated by the factory-scale robotics used by Cohen along with the introduction of YACS and the applications of PCR. Cohen's work also demonstrated that the whole genome can be mapped at low resolution without prior separation into individual chromosomes, as was originally envisioned. Generously, Cohen also made available to all researchers a library of thirty thousand cloned DNA fragments, representing 95 percent of the human genome. The success of Cohen's laboratory has led the United States to expand similar factorylike centers, and many laboratories have turned to whole genome mapping and the use of YACS has expanded.

Some worry exists about base sequencing methodology. Although the cost of sequencing has been reduced from about five dollars per bp to about one dollar, it needs to be halved again. More important, the speed of sequencing ought to increase by two orders of magnitude. It is not clear whether new ideas are available to break the logjam, but innovators are going to be encouraged. Similarly, the informatics system is barely keeping up with the submission, analysis, and distribution of data now, and it must speed its operations substantially to keep up with exponentially growing demands.

One shift of NIH strategy is a response to rapid progress in identifying genes and in cloning and partial sequencing of cDNA, which represents just the part of the DNA that codes for gene product. Ironically, the rush for cDNA research was given immense impetus by the NIH, which had insisted on the importance of sequencing the entire genome and not just the expressed genes. Craig Venter, then at the NIH, isolated large numbers of cDNA clones and sequenced the first 300 to 500 bases at high speeds, automating much of the work. NIH applied for broadly framed patents on (ultimately) 6,000 cDNA's.[35] The British filed for patents on 2,000 cDNA's[36] they had worked out to a similar extent. Strong protests[37] arose from every quarter that such incompletely identified entities were not patentable, and in the meantime, the applications would ruin the international cooperation and open sharing of information. After an initial rejection by the U.S. Patent and Trademark Office, the NIH decided to drop its appeal. In the meantime, it appears that many commercial concerns have been pushing ahead with such research at great speed, and Venter has set up a factory-scale, nonprofit center for doing this research, backed by commercial firms.[38] The current five-year plan reaffirms the importance of ultimately learning the entire structure of the genome, but recognizes a bias in the first several years to work on regions of biological interest. These regions will include cDNA, disease genes, and others where the gene function is known. Francis Collins, in establishing his laboratory at the NIH, sees urgency in gene identification: "The reason the public pays and is excited—well, disease genes are at the top of the list."[39]

Notes

1. U.S. Departments of Energy and of Health and Human Services, Understanding Our Genetic Inheritance, *The Human Genome Project: The First Five Years, FY 1991–1995*, NIH Publication no. 90-1590 (Washington, D.C.: Government Printing Office, 1990).

2. U.S. Department of Energy, *Human Genome 1991–1992, Program Report* (Washington, D.C.: Government Printing Office, 1992), 200.

3. Elise A. Rose, "Applications of the Polymerase Chain Reaction to Genome Analysis," *FASEB Journal* 5, no. 1 (January 1991): 46–54.

4. David T. Burke, George F. Carle, and Maynard Olson, "Cloning of Large Segments of Exogenous DNA into Yeast by Means of Artificial Chromosome Vectors," *Science* 236, no. 4803 (May 15, 1987): 806–12.

5. Ilya Chumakov, Phillipe Rigault, Sophie Guillou, Pierre Ougen, Alain Bil-
laut, Ghislaine Guasconi, Patricia Gervy, Isabelle Le Gall, Pascal Soularue,
Laurent Grinas, Lydie Bougueleret, Christine Bellaune-Chantelot, Bruno
Lacroix, Emmanuel Barillot, Phillipe Gesnouin, Stuart Pook, Guy Vaysseix,
Gerard Frelat, Annette Schmitz, Jean-Luc Sambucy, Assumpcio Bosch, Xavier
Estivill, Jean Weissenbach, Alain Vignal, Harold Riethman, David Cox,
David Patterson, Kathleen Gardiner, Masahira Hattori, Yoshiyuki Sakaki, Hi-
toshi Ichikawa, Misao Ohki, Denis Le Paslier, Roland Heilig, Stylianos An-
tonarakis, and Daniel Cohen, "Continuum of Overlapping Clones Spanning
the Entire Human Chromosome 21q," *Nature* 359, no. 6394 (October 1,
1992): 380–87.

6. Simon Foote, Douglas Vollrath, Adrienne Hilton, and David L. Page, "The
Human Y Chromosome: Overlapping DNA Clones Spanning the Euchro-
matic Region," *Science* 258, no. 5079 (October 2, 1992): 60–66.

7. Leslie Roberts, "NIH Takes New Tack on Gene Mapping," *Science* 258, no.
5088 (December 4, 1992): 1573.

8. Daniel Cohen, Ilya Chumakov, and Jean Weissenbach, "A First Generation
Physical Map of the Human Genome," *Nature* 366, no. 6456 (December 16,
1993): 698–701.

9. Lloyd Smith, "The Future of DNA Sequencing," *Science* 262, no. 5133 (October 22, 1993): 530–32.

10. The Huntington's Disease Collaborative Research Group, "A Novel Gene
Containing a Trinucleotide Repeat that Is Expanded and Unstable on Huntington's Disease Chromosome," *Cell* 72, no. 6 (March 26, 1993): 974–83.

11. Daniel R. Rosen, Teepu Siddique, David Patterson, Denise A. Figlewicz, Peter
Sapp, Afif Hentati, Deirde Donaldson, Jun Gote, Jeremiah P. O'Regan, Han-
Xiang Deng, Zohra Rahmani, Aldis Krizus, Diane McKenna-Yasek, Aunarue-
ber Cayabyab, Sandra M. Gaston, Ralph Berger, Rudolph Tanzi, John L.
Halperin, Brian Herzfeldt, Raymond Van den Bergh, Wu-Yen Hung, Thomas
Bird, Gung Deng, Donald W. Mulder, Celestine Smyth, Nigel G. Luing, Ed-
win Soriano, Margaret A. Pericak-Vance, Jonathan Haines, Guy A. Rouleau,
James S. Gusella, H. Robert Horvitz, and Robert H. Brown Jr., "Mutations in
the Cu/Zn Superoxide Dismutase Gene Are Associated with Familial Amy-
otrophic Lateral Sclerosis," *Nature* 362, no. 6415 (March 4,1993): 59–62.

12. Malek Djabali, Licia Selleri, Pauline Parry, Mark Bower, Bryan D. Young, and
Glen A. Evans, "A Trithorax Like Gene Is Interrupted by Chromosome 11q23
Translocations in Acute Leukemias," *Nature Genetics* 2, no. 2 (October 2,
1992): 113–18.

13. Gerard D. Schnellenberg, Thomas D. Bird, Ellen M. Wijsman, Harry T. Orr,
Leojian Anderson, Ellen Nemens, June A. White, Lori Binnycastle, James L.
Weber, M. Elisa Alonso, Huntington Potter, Leonard L. Heston, and George
M. Martin, "Genetic Linkage Evidence for a Familial Alzheimer's Disease Lo-
cus on Chromosome 14," *Science* 258, no. 5082 (October 23, 1992): 668–71.

14. Annemieke J. M. H. Verkerk, Maura Pieretti, James S. Sutcliffe, Ying-Hui Fu, Derek P. A. Kuhl, Antonio Pizzuti, Orly Reiner, Stephen Richards, Maureen F. Victoria, Fuping Zhang, Bert E. Eussen, Gert-Jan B. van Ommen, Lau A. Blonden, Gregory J. Riggins, Jane L. Chastain, Catherine B. Kunst, Hans Galjaard, C. Thomas Caskey, David L. Nelson, Ben A. Oostra, and Stephen T. Warren, "Identification of a Gene (FMR-1) Containing a CGG Repeat Coincident with a Breakpoint Cluster Region Exhibiting Length Variation in Fragile X Syndrome," *Cell* 65, no. 5 (May 31, 1991): 905–14.

15. Arthur H. M. Burghes, Cairine Logan, Xiuyuan Hu, Bonnie Befall, Ronald G. Worton, and Peter N. Ray, "A cDNA Clone from the Duchenne/Becker Muscular Dystrophy Gene," *Nature* 328, no. 6129 (July 30, 1987): 434–37.

16. John R. Riordan, Johanna M. Rommens, Bat-sheva Kerem, Noa Alos, Richard Rozmahel, Zbyszko Grzelczak, Julian Zielenski, Si Lok, Natasa Plavsic, Jia-Ling Chou, Mitchell L. Drumm, Michael C. Iannuzzi, Francis S. Collins, and Lap-Chee Tsui, "Identification of the Cystic Fibrosis Gene: Cloning and Characterization of Complementary DNA," *Science* 245, no. 4921 (September 8, 1989): 1066–73.

17. David N. Sheppard, Devra P. Rich, Lynda S. Ostedgaard, Richard J. Gregory, Alan E. Smith, and Michael J. Welsh, "Mutations in CFTR Associated with Mild-Disease-Form Cl-Channels with Altered Pore Properties," *Nature* 362, no. 6416 (March 11, 1993): 160–64.

18. Micheline Strand, Tomas A. Prolla, R. Michael Liskay, and Thomas D. Peters, "Destabilization of Tracts of Simple Repetitive DNA in Yeast by Mutations Affecting DNA Mismatch Repair," *Nature* 365, no. 6443 (September 16, 1993): 274–77.

19. Richard Strohman, "Ancient Genomes, Wise Bodies, Unhealthy People: Limits of a Genetic Paradigm in Biology and Medicine," California/Health Net Lecture Series, October 12, 1992.

20. Lap-Chee Tsui, "The Spectrum of Cystic Fibrosis Mutations," *Trends in Genetics* 8, no. 11 (November 1992): 392–98.

21. Delia Lakich, Haig H. Kazanzian Jr., Stylianos E. Antonarakos, and Jane Gitschier, "Inversions Disrupting the Factor VIII Gene Are a Common Cause of Severe Hemophilia A," *Nature Genetics* 5, no. 3 (November 1993): 236–41.

22. Jean Marx, "New Colon Cancer Gene Discovered," *Science* 260, no. 5109 (May 7, 1993): 751–52.

23. John Travis, "Biologists Visit New Orleans Under an Assumed Name," *Science* 260, no. 5105 (April 9, 1993): 162–63; Kenneth W. Kinzler and Bert Volgelstein, "A Gene for Neuroblastomatosis 2," *Nature* 363, no. 6429 (June 10, 1993): 495–96.

24. E. H. Corder, A. M. Saunders, W. J. Strittmatter, D. E. Schmechel, P. C. Gaskell, G. W. Small, A. D. Roses, J. L. Haures, and M. A. Pericak-Vance,

"Gene Dose of Apolipoprotein E Type 4 Allele and the Risk of Alzheimer's Disease in Late Onset Families," *Science* 261, no. 5123 (August 13, 1993): 921–23; Jean Marx, "Familial Alzheimer's Linked to Chromosome 14 Gene," *Science* 258, no. 5082 (October 23, 1992): 550; and Iain McIntosh, Ada Hamosh, and Harry C. Dietz, "No Linkage to Chromosome 14 in Swedish Alzheimer's Disease Families," *Nature Genetics* 4, no. 3 (July 1993): 218–19.

25. H. G. Brunner, M. Nelen, X. O. Breakfield, H. H. Ropers, and B. A. van Oost, "Abnormal Behavior Associated with a Point Mutation in the Structural Gene for Monoamine Oxidase A," *Science* 262, no. 5133 (October 22, 1993): 578–80; and Constance Holden, "Alcoholism Gene: Coming or Going?" *Science* 254, no. 5029 (October 11, 1991): 200.

26. Dean Hamer, Stella Hu, Victoria L. Magnuson, Nan Hu, and Angela M. L. Pattatucci, "A Linkage Between DNA Markers on the X Chromosome and Male Sexual Orientation," *Science* 261, no. 5119 (July 16, 1993): 321–27; Constance Holden, "NIH Kills Genes and Crime Grant," *Science* 260, no. 5108 (April 30, 1993): 619; and Eliot Marshall, "NIH Told to Reconsider Crime Meeting," *Science* 262, no. 5130 (October 1, 1993): 23.

27. Richard C. Mulligan "The Basic Science of Gene Therapy," *Science* 260, no. 5110 (May 1, 1993): 926–32.

28. John Maddox, "New Genetics Means No New Ethics," *Nature* 364, no. 6433 (July 8, 1993): 97; and Nelson A. Wivel and LeRoy Walters, "Germ-Line Gene Modification and Disease Prevention: Some Medical and Ethical Perspectives," *Science* 262, no. 5133 (October 22, 1993): 533–38.

29. W. French Anderson, "Human Gene Therapy," *Science* 256, no. 5058 (May 8, 1992): 808–13.

30. Diana W. Bianchi, Merton Bernfield, and David G. Nathan, "A Revived Opportunity for Fetal Research," *Nature* 363, no. 6424 (May 6, 1993): 12.

31. Christopher Anderson, "Genome Project Goes Commercial," *Science* 259, no. 5093 (January 15, 1993): 300–302; and Elizabeth Culotta, "New Startups Move In as Gene Therapy Goes Commercial," *Science* 260, no. 5110 (May 14, 1993): 914–15.

32. A. Dusty Miller, "Human Gene Therapy Comes of Age," *Nature* 357, no. 6378 (June 11, 1992): 455–60.

33. Francis Collins and David Galas, "A New Five-Year Plan for the U.S. Human Genome Project," *Science* 262, no. 5130 (October 1, 1993): 43–46.

34. Leslie Roberts, "Taking Stock of the Genome Project," *Science* 262, no. 5130 (October 1, 1993): 20–22.

35. Leslie Roberts, "Genome Patent Fight Erupts," *Science* 254, no. 5629 (October 11, 1991): 184–86.

36. Christopher Anderson, "UK Halts cDNA Patenting," *Science* 262, no. 5135 (November 5, 1993): 831; Christopher Anderson and Peter Aldous, "Secrecy

and the Bottom Line," *Nature* 354, no. 6349 (November 14, 1991): 90.

37. Editorial staff, "Gene Patents," *Nature* 359, no. 6394 (October 1, 1992): 348.

38. Christopher Anderson, "Controversial NIH Genome Researcher Leaves for New $70 Million Institute," *Nature* 358, no. 6382 (July 9, 1992): 92; Diane Gershon, "SmithKline Backs Sequencing Company," *Nature* 363, no. 6428 (June 3, 1993): 387.

39. Collins and Galas, "A New Five-Year Plan," 43–46.

[3]

The Human Genome Diversity Project: Ethical, Legal, and Social Issues

Henry T. Greely

OUR RAPIDLY INCREASING KNOWLEDGE of the human genome has led to great concern about the ethical, legal, and social implications of this knowledge. But there are 5.5 billion human genomes, not just one. The genetic variations between people can provide valuable evidence about human history and human disease; they also raise distinctive ethical questions. The Human Genome Diversity Project is a proposal to study the genetic variation within our species. The significance of human genetic diversity; the proposed project; and the special ethical, legal, and social issues the project raises are the subjects of this chapter.[1]

Human Genetic Diversity

The "human genome" is the term given the DNA carried in every human cell. In human beings, almost all of that DNA is carried in the forty-six chromosomes found within the cell's nucleus, autosomal chromosomes 1 through 22, and the two "sex" chromosomes, X and Y.[2] The chromosomes contain DNA with about three billion base pairs, the "letters" of the genetic code. Humans have an estimated 50,000 to 100,000 genes, each made up of letters from the genetic code. These genes give instructions for the construction of ("code for," in the jargon) proteins by the cells. But most human DNA does not seem to be part of any genes; more than 90 percent of it appears not to code for proteins or anything else. Its uses, if any, remain largely mysterious.

In almost all humans, almost all genes are almost identical. These genes have to be very similar, or else the bodies they build would not work and their owners would die. On average, the genetic sequences from any two humans will differ by only about one base pair in 1,000—999 will be identical. If one looks only at coding sequences, those that give instructions for making proteins, any two people will, on average, differ by only about one base pair in 10,000.

Some DNA bases and sequences can differ from person to person without changing anything because they do not seem to have any effect on how the body functions. Other variants produce such effects as variation in our height, eye color, fingerprints, blood groups, and whether we can roll our tongues. Sometimes, particular genetic variants can lead to susceptibility to disease or to unusual resistance to disease. Genes of these types exist in all human populations and are of great interest to medical researchers trying to improve human health and welfare. And even where the variations are of little, if any, functional significance, they can tell us something about the human past.

Variations are not distributed completely randomly. The more closely related two people are, the more likely they are to share the same variant genes, or alleles. Siblings have a 50 percent chance of sharing any one allele; less closely related family members have a smaller chance. To the extent that people who live in a particular area, or who make up an ethnic group, a nation, or a population share common ancestors, they are more likely to share these kinds of genetic variants than people who do not share common ancestors. Of course, ultimately, we all share the same ancestors; but for tens of thousands of years, for example, the ancestors of Native Americans did not live in the same region as (and hence did not have children with) the ancestors of Europeans. New genetic variants appeared in each group, and ancestral shared genes are present in different frequencies. Any one Native American may have the same alleles as any one European. When one looks at a number of alleles in a number of Native Americans and Europeans, however, it becomes clear that the two groups of people often differ in their variant genes. Thus, for example, the blood groups 0, A, B, and AB, all genetically determined, may appear in all human populations, but often in very different proportions reflecting the frequencies of the underlying alleles. Comparing the frequencies of different variations in different populations can reveal how recently they shared a large pool of common ancestors. Those frequencies can be used to see if, for example, the Irish are more closely related to the Spaniards or to the Swedes.

Two things are important to note about these differences in the frequencies of certain genetic variations among human populations. First, there are no particular genes that make a person Irish or Chinese or Zulu or Navajo. People in those populations are more likely to have some alleles in common, but no allele will be found in all members of one population and in no members of any other.[3] This cannot be very surprising in light of the vast extent of intermarriage among human populations now and throughout history and prehistory. There is no such thing as a genetically "pure" human population. *Ethnicity is a cultural label, not a genetic one.*

Second, although there are differences in the frequency of genetic variations among groups, the extent of such difference is small compared with the amount of difference found within a group. People within ethnic groups are genetically more different from each other than, on average, their group is from other groups. For an analogy, think about the size of professional football teams. The average weight of members of the San Francisco 49ers is probably not very different from that of the Dallas Cowboys, but the offensive linemen may outweigh the kicker by 150 pounds. Weight is more variable within the team than between the teams.

Even with those qualifications, this information can be valuable. First, it can clarify the major human migrations. It may show, for example, whether different migrations brought Native Americans to the Western Hemisphere from Asia or whether a single group is ancestral to all modern Native Americans. This research may make it possible to learn how people speaking Bantu languages expanded through much of sub-Saharan Africa in the last two millennia or how the Indo-European languages spread through Europe and Asia. It may determine who are the closest relatives to the Euskadi (also known as Basque) people of northern Spain and southern France, whose language seems unrelated to any other. And it may settle the continuing debate about whether *Homo sapiens* evolved to modern humans just once, in Africa, or several times on several continents.[4]

Studying human genetic variation will also provide increased knowledge about the factors that lead to disease or to health. People vary in their susceptibility to diseases, and although much of the explanation will be due to environmental factors such as diet, genetic predispositions play a role in many cases. The collection and analysis of DNA samples may, in conjunction with epidemiological evidence, help lead to the identification of genetic factors in some human diseases and eventually to ways to treat or prevent those diseases.

Finally, at a more fundamental level, without a broad study of human genetic variation, science will largely define the human genome, with its historical and medical implications, as that carried by the small number of individuals, largely of European ancestry, who are the subjects of the Human Genome Project. Broadening the scope of the exploration ensures that 90 percent of the world's population is not excluded from the human genome.

The Human Genome Diversity Project

The Human Genome Diversity Project was proposed as a way to include the whole of the world's population in what will become known as the human genome. Such a project was first suggested in a 1991 article by four distinguished geneticists and a historian of human genetic science.[5] Their proposal gathered momentum through a series of planning workshops in 1992 and 1993, culminating in an international meeting in Sardinia in September 1993. The Sardinian meeting formally established the project and defined its structure. Four months later, the international Human Genome Organization adopted the project.[6]

The project has an international executive committee, but its work is to be done by regional Human Genome Diversity Committees. Regions have established such committees in North America, Europe, South America, and China, with committees in formation in Australia, India, and other areas. The committees will be responsible for the operation of the project in their areas, subject to the guidance of the international executive committee. Only the European Committee has started any substantial work; for the most part, the project remains in its planning and fund-raising stages.

The project's goal is to advance the study of human genetic diversity. To do that, it wants to perform four tasks with human DNA samples: *collection*, *preservation*, *analysis*, and *database creation and management*.

The definition of separate human "populations" is not precise, but using language as the primary criterion, there are between four thousand and eight thousand distinct human populations around the world. As an interim goal, the project has suggested collecting DNA samples from five hundred of those populations within five years. Many samples may be collected from different parts of large linguistic or geographic populations, which cannot be adequately represented by a single sample. The DNA samples can be taken from small and harmless blood samples, from hair roots, from cells scraped off the inside of a cheek with a tongue depressor, and from sputum and other biological materials.

The project will preserve these samples in both central and regional repositories around the world. Most samples will be frozen; some blood samples will be changed into cell lines, which are capable of producing a large amount of duplicate DNA for study. For only the cost of making and shipping the duplicate samples, they will be made available to qualified scientists interested in doing research on them. Such availability or use of such samples may be limited by contracts between the project and researchers in order to protect the interests of the sampled populations.

The project itself plans to carry out basic, preliminary analyses of the DNA samples. It might subject all samples to the same set of tests on variation at the DNA level to provide uniform data across the collection.

Once the samples have been analyzed, as part of the standard battery funded by the project and by researchers who have received samples from the project, the results of the analyses will be put into a computerized database. Information in the database will be broadly available to those who want to use the results for legitimate research, again subject to contractual limitations.

Ethical, Legal, and Social Issues Raised by the Project

The project raises social questions every bit as fascinating and difficult as the scientific and historical questions it seeks to answer. Unless the project faces these questions directly and answers them well, they could destroy it. Issues arise in three areas: the collection of samples, the intellectual property treatment of the samples, and the possible consequences of the project for racism. For each area, this chapter sketches the nature of the problems, assesses their scope, and suggests possible responses.[7]

Collecting Issues

These issues arise from the fieldwork needed to collect DNA samples. The problems are largely ethical and legal, although they may often be political in the context of the population to be sampled. They have attracted the attention largely of bioethicists and of some groups representing indigenous peoples. The most important ethical issues in collecting revolve around informed consent, medical screening and services, and confidentiality.

Informed consent has proven to be a significant but difficult concept in both medicine and research. Its core concept—the recognition of the humanity and autonomy of the patients or research subjects—is widely accepted, but implementing that concept is often difficult. This is particularly true for this project, where the populations will often be isolated from the mainstream of Western culture, and the research involves particularly complicated scientific concepts. In addition, the project raises a related issue, common in anthropological work but not in biomedical research, of respect not just for the autonomy of the individual donor but for the integrity of the population's culture.

In recent years, there has been an interesting debate about the role of Western ideas of autonomy in seeking informed consent in other cultures. The use of Western ideas has been attacked as "ethical imperialism" and defended as upholding basic human rights. This debate, though fascinating, is moot to the extent that federal funds are used in the project. If such funds are used, the project must abide by federal law, which requires Western notions of individual information and personal consent. Reconciling culturally appropriate informed consent and legally required informed consent may be demanding, but the project will have to accomplish just that.

There is every reason to think that the project will succeed at that task. The informed consent problem, though tricky, is not new. Anthropological fieldwork with federal funding currently goes forward only after Institutional Review Boards have passed on the validity of the informed consent procedures. Although that fieldwork must be conducted in ways that vary according to the cultures of the populations studied, Institutional Review Boards have approved protocols that have been consistent with culturally appropriate informed consent. The project will have a higher profile than most anthropological fieldwork and its informed consent may be subject to greater scrutiny, but past methods should be adaptable to the legal and ethical requirements in this project. Particular attention will have to be paid to the novel issue of how to explain the processes and goals of this research in cultures far removed from modern biological knowledge.

But this project raises another novel informed consent issue: Whose consent is needed? In most research, consent is taken from the research subject and, where appropriate, his or her parents or guardians. For this project, the research subject is, in a very real sense, not only the person donating a sample but also his or her entire community. The project believes consent should come both from the individual participants and from their communities, through whatever body is most appropriate to give that consent. And the communities should be asked to decide only after receiving clear and understandable explanations of the project. It is hoped that this kind of involvement could carry over into a full partnership between the project scientists and the sampled community. At the very least, no group members will be sampled without informed consent from both them and their groups.

The medical issues are complicated. The project will be taking a potentially hazardous substance—blood—from its donors. Does it have an obligation to the donors, their community, or workers who will subsequently handle the blood to screen that blood for pathogens? If the project does have a policy about such screening, what information should be given to the donor and under what conditions? Does the project have an ethical obligation to provide treatment for conditions it discovers? To what extent can the project provide medical services to the populations sampled without, in effect, unethically "buying" the donations with those services? Or is the failure to provide medical services or other relevant quid pro quo in itself an ethical failure? These and related questions do not have obvious answers, but should be addressed collectively by the project.

Finally, confidentiality must be considered. In the process of collection, confidentiality about who donated samples and who did not may not be possible or culturally appropriate. Apart from possible medical uses, though, there seems to be little reason to link any given sample with an identified person.[8] Confidentiality can also be an issue for an entire population. It may not want its identity and location revealed, although that information could have scientific value for reviewers of the samples. There seems to be little reason not to provide as much confidentiality for individuals as possible and for populations as requested, but this issue must be considered in detail by the project before collecting begins.

These three areas do not exhaust the ethical and legal issues associated with collecting, but they may be the most important ones. The project should be able to deal adequately with them, just as researchers have dealt with the same issues in existing projects.

Intellectual Property Issues

The intellectual property status of the project's samples and data raises political problems, along with some very tricky legal issues. The underlying problem is simple: Who will get the profits from any commercial products stemming from this project?

The primary response of the project's organizers has been that there would not be any profits. The information being sought is unlikely to lead to new screening tests or pharmaceuticals. But upon close examination, no one has been able to say that such products cannot result from the project. In fact, one group concerned about this issue has pointed out that American researchers have sought a U.S. patent for a cell line produced from the blood of an indigenous woman from Panama and two indigenous people from Melanesia. And whatever the true prospects for commercial use, many people outside the project believe that such products will follow.

However real this issue may prove as the project unfolds, it is real now. Many developing countries have been outraged by what they view as Western exploitation of the genetic diversity of their plants and animals for profit. Western firms, they feel, take their potatoes, maize, or rare plants and use them to produce hybrid seeds or drugs. Far from giving profits of these variants back to the home country, the firms patent the results and sell the products back to their "source" at high prices.

Although this perspective on the international flow of genetic resources is not universally shared—especially not in the developed world—it is very strong in much of the developing world. It was the driving force behind the royalty provisions contained in the Rio Biodiversity Treaty, provisions that held up the adherence of the United States to that treaty. Scientists in developing countries, notably India and Kenya, have concluded that this issue would also be a major political problem for this project in their areas. And it has already been one focus for attacks on the project.

The response to this problem should be easy in principle, though it may prove very difficult to implement. At its organizing meeting in Sardinia, the project's organizers took the position that they did not seek any financial gain from this research. Instead, they resolved that the project should try to reserve and to protect the rights of the sampled populations by agreements regulating access to the project's samples. One option might be to say that no intellectual property rights could be claimed based on those samples without the express approval of the persons sampled. Another course would be to allow such development but require that royalties be paid on any product developed through the use of the project's samples or data. These royalties could be paid to an international organization, such as UNESCO, or a private foundation, for the benefit of the sampled populations.

Implementing such plans would be complicated. It may prove hard to bind users of the project's samples or, especially, the project's data. The choice of a "holder" for a royalty fund may prove contentious, as may be the decision about for whose benefit such a fund should be used. But for political reasons in dealing with a developing world already extremely sensitive to these issues in plants and animals, the project must have, from its beginning, a plan for avoiding any financial benefit to itself while reserving the benefits, if any, for the sampled populations.

Racism

This is largely an ethical and political question, with limited legal implications. The problem stems from humanity's sad—and continuing—history of group-defined hatreds and atrocities. While the geneticists involved in the project seem to agree uniformly that its expected results should undercut racism, everyone recognizes that data can, and usually will, be misused.

Consider, for example, a researcher's report that a particular allele of a certain gene is associated with higher levels of schizophrenia. Another researcher, using the project's data, may announce that the Irish (to use an example from part of the author's genetic heritage) have an unusually high incidence of this gene. Someone else, scientist or not (but probably English), then publishes the conclusion that "this proves the Irish are crazy." Such a result would be, at the very least, unpleasant. It takes little imagination to vary the names and traits in ways that would make the results tragic.

A historical example of exactly such misuse occurred between World Wars I and II. Much research was done on the frequencies of different ABO and rhesus blood types among different populations. It turned out that northwestern Europe had an unusually high incidence of type A blood and an unusually low incidence of type B blood. According to work done by historian William Schneider, the result, when turned into ratios and displayed on maps by Nazi ideologues, "demonstrates that Germany was the bastion of type A blood, holding out against the relentless onslaught of type B blood from the south and east."

It seems unlikely that the project would have much effect, one way or another, on such deep-seated human pathologies as racism and hypernationalism. No one claims that arguments based on blood types brought the Nazis to power or contributed in any appreciable way to their crimes. At most, it provided one more rationalization for conclusions already reached. But many persons involved in the project are repelled by the idea of providing even such limited support to racism. And in American politics, concerns about looking at genetic differences among "races" are understandably quite strong—particularly among African Americans. The fate of the 1992 conference at the University of Maryland on genetics and crime is some evidence of the strength of these concerns.[9]

How can the project deal with such misuse of its findings? Like any other scientific endeavor, it cannot completely prevent the misuse of its findings for any ends, noxious or otherwise. Nor should it seek to limit access to its samples or data based on the political views of the researchers. But the project can and should feel and act on an ethical obligation to put its data and findings in their proper scientific context. In the long run, that can be attempted by a program of public education about human genetic diversity. In the short run, the project could prepare ready response teams of scientists who would be available to the media to put such claims into context. It could even require short advance notice of publication of results based on its data in order to give it time to prepare such a response.

Conclusion

The social issues raised by the Human Genome Diversity Project are many and complicated. After much thought and many conversations about them, none of them seems so significant as to cast into doubt the value of proceeding with the project. On the other hand, they are real problems that must be confronted.

These issues, even more than sample size, collecting priorities, or funding, will make or break the project.

The project can succeed only if populations and individuals are willing to give it DNA samples. All the issues discussed in this chapter can become public relations problems among the possible donor populations. This is particularly true in light of the great fear and misunderstanding of genetics, even in supposedly "advanced" cultures. Already, press releases and E-mail postings have infected many with the unscientific fear that the project will promote biological warfare against indigenous peoples. Other rumors have apparently accused the project of seeking to "clone Indians and turn them into slaves."

These fears may seem ridiculous, but they are nonetheless real. If the project allows this kind of fear to run rampant and lets the social issues discussed here continue without resolution, it will fail through a lack of donors.

For these issues, the organized nature of the project is both a curse and an opportunity. It is a curse because its high profile will draw very close attention to these issues, attention that the current isolated efforts do not attract. It is an opportunity because an organized project can seek out solutions to these problems, guide investigators in applying the solutions to their situations and, to some extent, oversee the work to ensure that ethical and legal standards are being met.

Notes

1. The views expressed in this chapter are those of the author and, except as expressly noted, are not necessarily the views of the Human Genome Diversity Project. This chapter benefited enormously from comments from the author's colleagues on the North American Committee and from a seminar organized by the Wenner-Gren Foundation for Anthropological Research.

2. Organelles within the cell called mitochondria have their own DNA. They do have some particularly interesting uses in studying human genetics because they are passed down only from the mother. The size of the mitochondrial genome is trivial, however, compared with the genome found in the cell's nucleus.

3. There may be rare variations found only in some populations.

4. For a fascinating attempt to begin to tell the story of human migrations through genetic variations, see L. Luca Cavalli-Sforza, Paolo Menozzi, and Alberto Piazza, *The History and Geography of Human Genes* (Princeton, N.J.: Princeton University Press, 1994), or an earlier and more accessible version of his research, L. Luca Cavalli-Sforza, "Genes, People, and Language," *Scientific American* 265 (November 1991): 104.

5. L. Luca Cavalli-Sforza, A. C. Wilson, C. R. Cantor, R. M. Cook-Deegan, and

M.-C. King, "Call for a Worldwide Survey of Human Genetic Diversity: A Vanishing Opportunity for the Human Genome Project," *Genomics* 11 (1991): 490.

6. The Sardinian meeting produced a report that serves in some respects as a charter for the project. This document has not yet been formally published. "The Human Genome Diversity (HGD) Project: Summary Document," HGD Project Executive Committee, London, November 1994. The HGD project received approval in the United States by the National Research Council (NRC) and was recommended for funding on October 21, 1997.

7. The controversies caused by some of these issues are discussed in two articles: Patricia Kahn, "Genetic Diversity Project Tries Again," *Science* 266 (November 4, 1994): 720; and Lori Gutin, "End of the Rainbow," *Discover* 15 (November 1994): 70.

8. Such linkage may turn out to be necessary for kinship determinations or even to try to prevent, over time, duplicative donations.

9. The National Institutes of Health had funded a conference at the University of Maryland concerning genetic factors in crime. When the conference came under political assault as "racist," the NIH canceled the grant. See Philip J. Hilts, "U.S. Puts a Halt to Talks Tying Genes to Crime," *New York Times*, September 4, 1992, 1:1. The full saga of the conference, including an ordered reconsideration of the cancellation of the grant, can be followed in a series of articles in *Science*: Joseph Palca, "NIH Wrestles with Furor Over Conference, *Science* 257 (August 7, 1992): 739; Constance Holden, "Back to the Drawing Board, Says NIH," *Science* 257 (September 11, 1992): 1474; Richard Stone, "HHS 'Violence Initiative' Caught in a Crossfire," *Science* 258 (October 9, 1992): 212; Eliot Marshall, "NIH Told to Reconsider Crime Meeting," *Science* 262, no. 5130 (October 1, 1993): 23; Charles C. Mann, "War of Words Continues in Violence Research," *Science* 263 (March 11, 1994): 1375.

[4]

Fair Shares: Is the Genome Project Just?

Karen Lebacqz

PATRICIA KING, PROFESSOR OF law at Georgetown University and one of the few African Americans to testify before the Senate and House hearings of 1989 and 1990,[1] declared, "What the genome initiative will do is to make absolutely clear—if we are not convinced already—that we have basic concepts and values on a collision course."[2] The Human Genome Project (HGP) incorporates fundamental tensions around certain values. One of these values is justice or fairness.

The history of the HGP is rife with arguments about fair shares. In the Senate and House hearings of 1989 and 1990, one cluster of arguments about fair shares focused on national interests and international justice.[3] This cluster focused on fairness in three areas: fairness in burden sharing, fairness in benefit sharing, and fairness in profit sharing. I will present those arguments as they emerged, elicit the understanding of "fair shares" that appears to be embedded in them, identify the concept of justice implied by that understanding, and offer some preliminary ethical critique.[4] I will also suggest a contrast between the philosophical view of justice that dominated the arguments about fair shares in the hearings and a possible theological understanding of justice and fair shares.

The Arguments

Arguments made during the hearings about fair shares and international competition clustered around three issues.

The first of these was signaled by Hon. Ralph M. Hall. He opened the House hearings of October 1989 by asking, "How does the U.S. make sure that the basic research effort . . . is shared equitably among our international scientific partners while recognizing that the nation who leads the applications resulting . . . will have a competitive advantage in pharmaceutical biotechnology and related industries?"[5] The first issue posed was how to ensure that all countries bear a fair share of the burdens of research. The issue, then, is the *fairness of burden sharing*.

The second issue was hinted at by Hall and expressed bluntly by several in the 1989 Senate hearings. Senator Hollings queried, "How can we guarantee that American firms get the first crack at the technology that the Human Genome Initiative will provide, so as to stay ahead of our foreign competition?"[6] Senator Bryan declared, "The bottom line is that we want our team, so to speak, to get any of the benefits of commercial applications arising from major breakthroughs."[7] The goal here is not fairness in the sharing of burdens, but *fairness in the reaping of benefits*.

A related concern was given voice by Representative Packard: "We must protect the U.S. taxpayer's investment in this new big science project."[8] Although taxpayers were not mentioned frequently, their presumed interests formed an unwritten testament in the background. Since government monies are largely raised by taxing citizens, and since the allocation of government monies was at stake, support for the HGP could be seen also as a matter of ensuring a *fair share of return on investment* to taxpayers.

These three arguments are closely linked. "Who should bear the burdens of scientific research," "who should reap the benefits," and "how can those who invest in research get a fair share of the returns" are parts of a whole. They have to do with the justice of allocating burdens and benefits.

Precisely because they are so closely linked, assumptions about one affect assumptions about the others. The arguments were lumped together in the hearings, as one would expect them to be. Nonetheless, it is possible to pull out some of the background assumptions and implicit arguments addressed to each of these concerns.

Background Assumptions: The Role of Biotechnology

Underlying concerns for burden sharing, competition, and return on investment was recognition of the economic impact of the biotechnology industry:

"The biotechnology industry that will grow from this . . . is going to be enormous."[9] Data were offered to indicate just how enormous the biotechnology industry might be: "The market value of American biotechnological products is already over $1 billion per year and is expected to reach as much as $40 billion per year early in the next century."[10]

Because of the potential financial impact of the biotechnology industry, concern for technology transfer was built in to the HGP from the beginning. The project was presented to Congress not solely as a matter of basic scientific research, nor even for its medical potential, but as a boon to American industry. James Watson, then director of the National Center for Human Genome Research, was a major figure in the hearings and probably the key scientific spokesperson for convincing the House and Senate to support the HGP. Watson declared at the outset: "We consider the involvement of industry in the genome project to be crucial to its success. The applications that will derive from the basic information will be developed in the private sector for the most part."[11]

Indeed, Watson indicated the intention of the National Institutes of Health (NIH) to give grants not only to universities but also to private industry for some of the genome work.[12] "What is good for General Motors is good for the United States, or vice versa," he declared. "Our program will be in trouble if our companies are in trouble."[13] The project was presented as good for the American economy. It was also presented as requiring the involvement of industry from the very beginning.

Behind a concern for market value and for the involvement of American industry was a concern for the balance of trade. Biotechnology is one of the few areas at present where the United States has a positive balance of trade: "Biotechnology offers America a real opportunity to redress our failure in industrial competitiveness and, accordingly, the negative trade balance."[14]

The target of concern about balance of trade was—and still is[15]—Japan. Leroy Hood signaled the concern when he testified: "America is currently the world leader in biotechnology. This leadership is unequivocally being threatened by the Japanese."[16] Much of the discussion about "fair shares" arose, therefore, in discussing the role of Japanese research and industry versus U.S. research and industry.

The concern is simple. On the one hand, Japan is perceived as not putting much effort into the basic research sector: "On the whole, the Japanese have a very badly funded basic science program."[17] On the other hand, they are perceived as excelling at technology transfer. The fear was that Japan would not support the basic research of the HGP, but would be the first to capitalize on commercial applications: "They would let us pay for the basic research and they'd spend their money on commercial applications and pursuits."[18]

The tone of international competition therefore permeated the hearings. Senator Domenici, calling himself a "true believer" in the human genome,[19] reviewed the history of efforts at legislation on biotechnology research. The 1987 Department of Energy proposal was called the Cooperative Research Initiatives Act. It was incorporated into a broader biotechnology bill in 1988 under the title Biotechnology Competitiveness Act. The heart of that proposal was then incorporated into the National Competitiveness Technology Transfer Act of 1989.[20] This legislative history indicates how "cooperation" was transmuted into "competitiveness" as the links between research and biotechnology transfer became more prominent.

In such a competitive atmosphere, fair shares were taken to mean that all countries must bear their part of the burdens of research and—at the same time—that provision must be made so that U.S. industry would not be disadvantaged in technology transfer. A number of themes emerged: fairness as equal opportunity, fairness as getting rewards where one has carried burdens, fairness as sharing of basic goods. These themes are potentially in tension, as Patricia King suggested.

Sharing Burdens

The key argument on sharing of burdens and benefits comes from James Watson: "Sharing data without sharing costs strikes me as very unfair."[21] The basic notion of fair shares is that those who would reap benefits must also share burdens. For one country to pay for basic research while another reaps the rewards of industrial and commercial applications was considered unfair. If Japan or other countries are to share in the benefits deriving from the genome research, then they must also share in the burdens of supporting that research.

Watson went so far as to suggest that if the Japanese government did not undertake to support basic research more adequately, "we have no choice" but to regard Japan as an enemy.[22] Even James Wyngaarden (then associate director for Life Sciences of the Office of Science and Technology Policy in the Executive Office), who supported international collaboration and free exchange of information, urged that genuine collaboration had to be a two-way street and not subsidization.[23] Thus, coupled with the concern for an American competitive edge was a notion of fair shares that would require Japan and other countries to contribute to the basic genome research. Anything else would constitute not "fair shares," but an unfair advantage for certain countries. Only by requiring countries to share in the research effort would *burdens* be distributed equitably.

One way to ensure that all countries participate in the basic research effort is to refuse to share information with countries that do not. Thus, any country not carrying some of the burdens of basic research would also not be given the information from which commercial benefits might be derived. That this was in fact proposed becomes clear when we look at the issue of benefits.

Securing Benefits

The equitable distribution of burdens does not automatically ensure the equitable distribution of benefits. If the concern was partly to ensure international support for genome research, it was primarily to ensure that U.S. industry and science got the jump on the development of new technologies. The desire to retain a competitive edge requires more than simply sharing the burdens of research.

The primary mechanism proposed to ensure the competitive edge was patenting, secrecy, or nonsharing of information at early stages. The notion of fair shares, then, was taken by most participants to include some proprietary rights to information and some arenas in which sharing was not required. In his written responses to questions put to him following the House hearings of 1989, for example, Watson noted that "collaborators traditionally receive information quite a while before the general scientific public does."[24] He further argued that "it will be important for companies to obtain international patent protection for inventions arising out of the genome project."[25]

Although Watson did not make a precise proposal in that setting, his arguments imply a three-level process of sharing information. On the first level, those who develop information (e.g., isolate a sequence of nucleotides) should have some time to try to discern its meaning before publishing or making that data available to others. This stage of secrecy would include U.S. researchers and any international collaborators. The term "collaborators" was never clearly defined in the hearings, but presumably means those in laboratories both here and in other countries who are working in close consultation and with an agreement about collaboration, such as the arrangement between the University of Michigan and the Children's Hospital in Toronto that share credit for the discovery of the gene for cystic fibrosis.

Information would then be shared with scientists in other countries who have made a commitment to basic genome research and are carrying their fair share of the costs of that research generally, though they might not have been involved in the specific research where a breakthrough has occurred. Watson originally proposed that genes might be divided among countries. If that were done, then countries not working on a specific gene but nonetheless contributing to basic genome research would all get the information at the same time.

Finally, the information would be published and everyone would have access to it, including scientists in countries not supporting basic genome research.[26]

This implied proposal is similar to common practice today: research teams, including collaborators in different countries, who make a breakthrough may hold their information for a while, but as soon as it is published, it becomes available to the general scientific community. However, Watson's implied proposal adds another step: between the secrecy of the team and general sharing with the entire scientific community, there might be sharing with those in countries supporting basic genome research.

Maynard Olson, professor of genetics at the University of Washington, agreed with Watson in urging that the laboratory that initiates the finding should be given time to analyze it before it is published.[27] Once that has occurred, however, Olson urged that data then be made available to the remainder of the scientific community "regardless of nationality."[28] Thus, where Watson appeared to suggest that some distinction be made between countries that have supported the basic research effort and those that have not, Olson proposed that all countries have access at the same time. His would be a two-step rather than a three-step process, closer to common practice today.

Underlying these implied proposals are several issues about fair shares. It seems to be agreed that it is fair for those who have made a discovery to have time to capitalize on that discovery. This is a form of assuming that those who have borne certain costs should reap any rewards associated with the costs. The question, then, is whether those who have borne general costs (though not the specific costs of the particular research) should also reap the benefits earlier than would the remaining scientific community.

The link between scientific research and U.S. industry is also crucial to understanding how the burden-benefit equation would give U.S. businesses a jump on other countries. The time given to scientists to develop the implications of their discoveries benefits business only if there are close links between business and the scientists doing basic research. While most basic research is still done in universities, the increasing tendency for university departments to contract with corporations also suggests an easier avenue for the burdens of basic research to be translated into commercial benefits.[29] As noted above, support for commercial links was built into the HGP from the beginning.

Getting a Return on Investment

Very little was said about taxpayers and what would constitute a fair return for their investment. To a large extent, the answer to this question seems to be tied up with the answer to the question of fair burdens and benefits. The argument seems to be much the same: those who have invested (whether it is time, money, or talent) should reap the rewards before the rewards go to others. In light of the heavy emphasis on U.S. business and commercial concerns, it seems to have been the assumption that the return on the investment of taxpayers' dollars comes in the form of the jump that U.S. businesses would get on commercial applications arising from the HGP.

Fair Shares

In short, the conversations about fair shares in the HGP appeared to assume that those who bear burdens should reap the benefits that might be associated with the burdens, and that it would be unfair for any to reap benefits without having shared in the burdens. Burden bearers include taxpayers whose money supports basic research in the United States, scientists who invest their talents in the work, and—on a larger scale—entire nations whose policies and monetary contributions support basic science. All these forms of carrying burdens are understood to qualify one for benefits. We turn now to see what theory of justice underlies this approach to the distribution of goods.

An Underlying Theory

In *The Grammar of Justice*, Elizabeth Wolgast argues that the meaning of justice depends on the language that we use to talk about it.[30] Children learn about justice from the way their parents and peers talk about it: "share with Tommy," "she's smaller than you are, so she gets three tries instead of two," or "that wouldn't be fair to your brother." The word "justice" need not be used; the concept is conveyed by the language and tone. It is also conveyed by stories that children are told or read.

One such story came immediately to my mind as I read the government hearings, particularly the sections dealing with fair shares in international relations. The story concerns a little red hen who found a few grains of wheat.[31] She asked her friends whether they would help her plant the wheat. They declined, so she planted it herself. She asked whether they would help her water and tend the wheat. They declined, so she did it herself. She asked whether they would help her harvest the wheat, thresh it, and grind it. They declined, so she did it herself. She asked whether they would help bake it into a loaf of bread. They declined, so she did it herself. Then she asked whether they would help her eat it. They accepted with alacrity, but the little red hen said, "No, you won't! I found the wheat and I planted it. I watched the wheat grow and when it was time I harvested it and threshed it and took it to the mill to be ground into flour, and at last I've baked this lovely loaf of bread. Now I'm going to eat it myself." And she did.

This story was read to me as a child and is still read to children today in the United States. While it might be interpreted in many ways, the modern jacket captures a meaning reminiscent of the discussion of fair shares and international competition: "The resourceful biddy shows her lazy neighbors the value of honest work." The story is taken to mean that those who have not participated in the work effort from the beginning have no right to share in the rewards at the end. The one who works has proprietary rights over the product of the labor and need not share it if she chooses not to.

Such a notion appears to undergird the sentiment expressed toward Japan in the hearings on the HGP: if the Japanese do not participate in the basic research from the beginning, then they have no right to share in the rewards at the end. Just as the friends refused to help the little red hen and therefore had no fair share of the bread, so any nation that does not labor in basic genome research effort has no fair share of the "bread" of commercial applications.

This story and the notions of fair shares that emerged in the government hearings reflect a Lockean view of property, ownership, and the distribution of burdens and benefits. The Lockean theory argues that I can acquire "ownership" by mixing my labor with natural resources.[32] Something becomes "mine" rather than simply available for general use when I have taken the raw materials and put labor into them. Hence, the little red hen comes to "own" the bread because she has planted the seeds, watered the plants, harvested the wheat, ground the flour, and baked the bread. Her hard work gives her proprietary rights over the final product.

The arguments about international shares have something of the flavor of this argument. When Watson says, "Sharing data without sharing costs strikes me as very unfair,"[33] he gives voice to this basic Lockean understanding that only those who bear the burdens should reap the rewards. Japan or other countries that do not share the costs of genome research are akin to the friends of the little red hen who would not share the burdens of planting, watering, harvesting, grinding, and baking. Because they have not shared the work or the costs, it would be unfair for them to share the benefits or data.

Crucial to the Lockean theory is the notion that the original resources are "common" or open to all. Thus, proprietary rights depend on the notion that all have an *equal opportunity* to "mix their labor" with the raw resources. Equal opportunity at the outset is implied by the theory. Given equal opportunity, the product of labor belongs to the one who labors, and cannot be justly used by another without permission of—or payment to—the first.

Similarly, arguments about international competition in the genome effort assume that the genome is available for study to any who are willing to "mix their labor" with it. Once that labor is mixed, however, the products of the labor belong to those who have put their efforts into it, and not to those who stood by watching. They have right of access only through the mechanisms of exchange—for example, by purchasing diagnostic tests from companies holding the patents on those tests.

There is in Locke's theory an important proviso: the right of proprietary claim does not extend infinitely. One may garner not all the acorns, but only as many as one can use. Whatever is beyond this, suggests Locke, is more than one's share and belongs to others. The appropriation of land through labor was acceptable to Locke because he assumed that "there was still enough and as good left" for others.[34] Locke assumed that "no man's labor could subdue or appropriate all, nor could his enjoyment consume more than a small part."[35] Thus, Locke's understanding of the right to acquire ownership appears to stop short of ownership that would deprive others. One cannot claim ownership of the last water hole simply because one puts stepping-stones around its edge, thereby mixing one's labor with it. Appropriation is acceptable only where it does not make others worse off. Under some circumstances both equal opportunity and the right to acquire proprietary interests must yield to the common good. Sharing sometimes takes precedence.

The best known contemporary Lockean theory is Robert Nozick's *Anarchy, State, and Utopia*. Where Locke focused on original appropriation of land in particular, Nozick focuses on acquisition by transfer. "If the [Lockean] proviso," he argues, "excludes someone's appropriating all the drinkable water in the world, it also excludes his purchasing it all."[36] Each owner's holding, therefore, includes the "historical shadow" of the Lockean proviso: appropriation or proprietary rights may not worsen the situation of others. "A person may not appropriate the only water hole in the desert and charge what he will."[37]

Nonetheless, and significant for our purposes, Nozick also interprets the historical shadow of the Lockean proviso to mean that an invention that is not shared does not violate the Lockean proviso because it does not worsen the situation of others. A medical researcher who synthesizes a new substance that would effectively treat a disease and who refuses to sell it except on his or her terms has not made others worse off than they were before.[38] Nozick suggests that this approach may also illumine the situation of patents. An inventor's patent does not deprive others because the object of the patent would not exist without the inventor. Thus, no one is worse off than before. However, there is always the possibility that someone else might eventually invent the same thing. An unlimited patent would worsen that person's situation by preventing him or her from ever being able to "mix labor" with the resource and acquire proprietary rights. Hence, patents do not violate the Lockean proviso as long as they are limited.[39]

This, then, is the philosophical foundation for the notion that proprietary rights are acquired by those who mix their labor with natural resources, and that they may patent their inventions and need not share with others except under certain limited conditions. It is the foundation both for assuming that those who would share in the rewards must share in the labor and for assuming that it is permissible to patent what one has created through one's labor.

Critique

There are a number of problems with the specific arguments made for fair shares and with the Lockean understanding on which the arguments appear to rest.

Sharing Burdens

The basic argument about sharing burdens was that all countries—especially those that might be positioned to reap rewards from the genome research—should share in the burdens of that research. It would be unfair for the costs to be borne by the United States while Japan or others got the benefits.

Background assumptions were operating here. Key among them was the central contention voiced by James Watson that Japan was not contributing "its share" to genome research. In an agreement signed in November 1986, the Japanese government had in fact acknowledged that it had undersupported basic research.[40] While it took steps to correct this deficiency—for example, by opening four new positions for foreign scientists on its faculties—nonetheless its admission gives evidence to support the basic sentiment of U.S. scientists that Japan was not carrying its fair share of basic research burdens.

Notwithstanding this admission, the Japanese government may not have been as remiss in its support for basic science as Watson and others took it to be. The Japanese Ministry of Education, Science, and Culture (commonly known as Monbusho) gave roughly the equivalent of $2 million each year to the Japanese genome effort in 1989 and 1990. By U.S. standards this may not have seemed a large amount. In the United States, about 12.5 percent of research and development expenditures were going into basic research at that time, while the comparable expenditure in Japan was only 3 percent.[41] Thus, if "fair shares" requires an equivalent percentage from each country, the charge that Japan was not contributing its fair share would be correct. However, there are mitigating considerations.

First, Japan initiated a program focused on DNA sequencing in 1981, long before the Human Genome Project was proposed to the U.S. Congress. Charles DeLisi notes: "In 1985 . . . when we started developing the Human Genome Project . . . it was picked up by the American press as a new and bold initiative. In fact, it was not at all new in Japan. . . . My old colleague, Professor Wada . . . had already done what we were just beginning to think about."[42] Wada visited the United States in 1986, trying to forge international links and support for a genome effort. While Japanese scientists were seeking genuine collaboration, "loose historical analogies" about electronics and automobiles were used to convince members of Congress that Japan was a threat, not an ally, causing one commentator to suggest that "the Japanese genome project got stuck in the tarbaby of U.S.-Japan trade tensions."[43] Thus, the Japanese scientists had pioneered in some genome work and had sought a genuinely collaborative effort with the United States.

Second, the infrastructure in Japan is different from that in the United States. In the United States, a large percentage of basic research is supported by the government; in Japan, support for research comes largely from industry. This affects the focus of the research and results in the Japanese stress on applied or commercializable research. But it also means that the Japanese contribution to research cannot be measured simply by looking at the percentage of the federal budget that supports research. To do so is to impose U.S. ethnocentric standards. Considered by Japanese standards, the contribution to the genome project may have been a significant government commitment to a pilot project.[44]

Thus, the basic scientific effort on the genome was initiated early in Japan, and the Japanese government had made a significant commitment to that effort, understood from within cultural boundaries. The judgment that Japan was failing to contribute its fair share needs nuancing.

Moreover, Martin Kenney argues that the Japanese enterprise is "more highly structured and consequently less redundant" than the somewhat "chaotic" approach in the United States.[45] If the provision of a better infrastructure for industry in Japan makes it better positioned than U.S. industry to capitalize on technology transfer, then it is not clear that there is any unfairness in Japan's superiority in technology transfer. If equal opportunity to develop resources is the standard for fair shares, then it would seem that where basic knowledge is given to both U.S. and Japanese industry at the same time, and Japanese industry capitalizes on it better than U.S. industry does or commercializes it first, there is no unfairness.

When the flour is ground, if the little red hen cannot figure out how to make bread of it and she shows it to her friends and one of them succeeds in making bread, was it unfair to the little red hen? As Kaname Ikeda, science counselor at the Japanese embassy in Washington, D.C., put it, "the same research is open to U.S. companies, too. Maybe the root of the problem is that Japanese companies are quick to commercialize findings."[46]

Securing Benefits

Ikeda's statement points to an interesting irony in the U.S. position: the goal of ensuring that all countries share in basic research and the goal of securing a competitive edge in applications for the United States may not be compatible. If all countries share fairly in the research effort, then all countries would and should have an equal chance at the commercial applications resulting. The only way to ensure the first jump at commercial applications may be to excel in the basic research. Thus, the desire to retain a competitive edge for the United States might precisely depend upon bearing more of the weight of basic research. Perhaps because the United States has excelled in basic research, American industries have been able to dominate in biotechnology.

Nonetheless, the proposed solution to securing benefits was patenting, secrecy, or nonsharing of information at early stages. (The little red hen could patent the ground flour and prevent others from using it until she figures out how to make bread.) Several concerns were raised within the hearings about this solution.

Although Maynard Olson agreed with the basic perspective that scientists who make a discovery should have time to develop it, he cautioned that secrecy is not always possible. As desirous as it may be to allow a laboratory time for its own work prior to international notification, this is not always feasible: "Francis Collins was unable to maintain even the normal prepublication embargo over the very exciting news from their laboratory [the discovery of the gene for cystic fibrosis]."[47] Olson noted that with increased collaboration, "too many people are involved" to be able to keep secrecy well. He therefore concluded that "it would be a catastrophe" to attempt to create secrecy. For practical reasons, the secrecy needed to safeguard the reaping of rewards is not always possible.

Olson's caution points to an interesting irony: arguments for government support of the HGP depended to a large extent on its economic implications for U.S. business. But those economic implications themselves depended on safeguarding the secrecy and patent rights of information developed. Without such patent rights and secrecy, there is the clear threat that Japan or other countries would take the basic discoveries and develop the commercial potential more quickly than U.S. industry would. Secrecy was considered necessary to ensure the economic gains.

If secrecy is not possible, the entire argument for economic gains might collapse. With it, the argument for support of the HGP might collapse. Olson concluded, "We would be misleading this committee if we told them that what they were buying and paying for the human genome initiative was some kind of proprietary American database that we would have a really substantial first crack at to skim the commercially important cream off of, or, indeed, the basic scientific discoveries."[48] Evidence from the hearings suggests that members of Congress thought they were buying a proprietary database from which U.S. industry would reap the rewards, contrary to Olson's caution.

Theoretical as well as practical arguments were raised against secrecy. George F. Cahill of the Howard Hughes Medical Institute and treasurer of the Human Genome Organization (HUGO) offered a perspective that seems to militate against notions of patenting or secrecy. "It is totally inconceivable that such an activity would be done without an open, active, and coordinated international effort," he urged.[49] Cahill's reasoning was not merely that it is impractical to do the genome project without international cooperation. He did note that one cannot study the American genome without attending to the fact that "we are all mixtures, except for the Native Americans,"[50] and hence must coordinate with studies of other genomes in order to understand our own. As important as these pragmatic reasons were, Cahill's basic argument was not pragmatic but philosophical. "The human genome belongs to the entire human race," he declared.[51] If it belongs to all, then it should be shared with all.

Hon. Harris W. Fawell of Illinois also argued that if the breakthroughs and benefits were as great as they had been argued to be, then "we ought to be able to take the attitude somehow that when the breakthrough comes, everybody immediately shares in that knowledge."[52]

Both Cahill and Fawell point to an *ideal* of free sharing based on an understanding of the importance of the genome research and on the conviction that anything so important cannot be claimed as proprietary territory. (Bread is so crucial for sustaining life that the little red hen should not be allowed to delay the process of discovering it or to garner all of it.) The ideal points to the notion that the human genome is not the property of some. It belongs to all; therefore, fair shares would allow all to participate in the rewards. This is akin to the Lockean proviso that prohibits proprietary claims on scarce resources where exercising the claim would worsen the situation of others.

At stake is the very question of ownership and proprietary rights. Does mixing one's labor with something give one the right of ownership, as Locke asserted? Is the genome a resource that cannot be appropriated without making others worse off?[53] Or is the discovery of genes and base pairs akin to an invention that can be patented because doing so does not make others worse off than they were before?

The question of whether the genome can be "owned" or whether one can hold title to nucleotide sequences has come up in vociferous debates over patenting.[54] It goes beyond the immediate scope of this chapter, but it has implications for assessing the nature of fair shares. If there are proprietary rights to the genome, then securing benefits from genome research will take one form. If there are not, then fair shares will be determined differently.

This question is of particular importance for so-called Third World countries that are not able to contribute to the basic research. Are they to be deprived of any benefits—or forced to buy those benefits from companies in the United States and other First World countries—because they are not able to bear a share of the burdens? A written question put to James Watson asked what would be the role of the developing countries. Watson replied that their major concerns at present are in the areas of training for their scientists and access to data and new technology generated by the HGP. "These interests are reasonable," he suggested, "and are consistent with the goal of making the basic data developed by the genome project freely available."[55] Watson's response hints that some other countries that are not able to contribute to the basic research should nonetheless have free access to data at some stage, suggesting that perhaps fairness in burden sharing must also take account of the financial resources of countries. Nothing was said in the hearings to address the question of fairness to Third World countries in terms of access to benefits of diagnostic tests or disease intervention.

Return on Investment

Representative Packard seemed to assume that fair shares would imply some proportional return on taxpayers' investment. If U.S. taxpayers support genome research, then they should reap the rewards. Members of the public who have carried the financial burdens of investing in basic research should reap the rewards that come out of that research.

Three questions go unspoken here, however. First is a question akin to that raised about developing countries. What about residents and citizens of the United States who are too poor to pay taxes? Are the taxpayers the only ones who can be considered to have "invested" in the HGP? Surely, the assumption is that the whole society will benefit from this research and from the boost that it will give to industry. Indeed, if the assumption is that benefit to U.S. business constitutes the return, then some of those who did not invest—for example, the unemployed—may reap the largest rewards, particularly if new jobs are created.[56] Clearly, the assumption is that all taxpayers then benefit, for more taxpayers are added to the rolls. That is, the fair shares for some cannot be calculated without assessing the entire web of the society. Fair shares to investors depend on the structures of the whole and on the common good. This raises immediate questions about the adequacy of a Lockean notion.

Second is the nature of investment. Most investments involve risk. Return is not necessarily guaranteed, particularly when the investment is in an enterprise such as basic science that may take years without yielding practical applications.[57] One could argue that "fairness" of return on investment means that one takes one's risks; if there are returns, investors will share them, but if there are not, they will share the losses. While one always hopes for a return from investment, it is never guaranteed. Whether there will be a practical return at all and what the *amount* of return will be are always a risk.

The third question is the *nature* of the return. What would constitute a return on taxpayers' investment? Are practical applications for U.S. business the important return? Packard had in mind garnering the "enormous economic potential" of the project for the U.S. citizens who support so much scientific research here. He also had in mind that the garnering of this economic potential—the safeguarding of the investment of taxpayers—had to do with making sure that U.S. industry got first crack at the applications of the genome research. Thus, the return would be seen in concrete economic terms, through the betterment of U.S. If "what's good for General Motors is good for the USA," then taxpayers can be presumed to reap returns on their investment when U.S. businesses thrive.

This is not the only way to interpret the interests of or return to taxpayers, however. Martin Kenney argues that public support for scientific research is in large part an investment in an open university system that upholds values such as free exchange of ideas and collaborative research. He further argues that in an open university system, the university is not perceived as being a for-profit enterprise.[58]

Kenney did not participate in the hearings, but he might have argued that the "return" expected by U.S. taxpayers comes in the form of support for the university research system as it has existed with all of the benefits that accrue from it. When that system is altered by the exigencies of the business world, taxpayers, far from gaining a return on investment, have lost the most important return: the university is transformed from its fundamental role of producing for the general society to a function akin to being a research team "leased" by a corporation.[59] The university and its professors lose their neutrality and independence when business interests dominate. In the long run, this could have disastrous consequences: "If the universities become the captives of chemical companies and high technology firms, public trust in academia will rapidly disappear."[60]

In other words, fair shares require an examination of the basic structure of scientific research in this country. Kenney's argument is that the changing laws that encourage the university-industrial links that are so central to the HGP in fact create new social relationships within the university and between the university and society. These new social relationships undermine the traditional role of the university and may be a large price to pay for the investment in research. Although Kenney is not writing specifically about the HGP, his concerns for biotechnology generally suggest that he would argue that the public will not have received a "fair" return on investment, but will have bought a Trojan horse, if support for the HGP is bought at the price of secrecy and shifting fundamental patterns of sharing research.

In short, the interest of taxpayers is not simply the good of U.S. industry. Taxpayers also have an interest in the educational system and in the values that it supports. Testifying earlier, Jonathan King argued: "The openness, the free exchange of ideas and information, the free exchange of strains, of proteins, of techniques, have been a critical component in the creativity and productivity of the biomedical research community."[61] If openness and free exchange of ideas are the values that U.S. taxpayers want to support, then they will not get a fair return for their investment if these values are threatened.

It would take an entire chapter to address the importance of openness and free exchange to the scientific endeavor. One might point out, however, that Watson and Crick could not have discovered the structure of DNA without two crucial pieces of information, one from Linus Pauling and one from Rosalind Franklin.[62] The first was given to Watson and Crick not by Pauling himself, but by his son, Peter, who happened to share an office at Cambridge with Watson and Crick. The second was given to Watson and Crick not by Franklin herself, but by colleagues.[63] Had secrecy prevailed, neither Pauling nor Franklin might have shared with Watson and Crick, and the latter would not have obtained two pieces of information crucial to their discovery.

Indeed, Watson in 1971 accused an NIH researcher of withholding a viral strain from Watson's Cold Spring Harbor laboratory, saying that "public money was being spent to develop reagents not being made available to other people." Here, Watson joins the chorus of those urging free exchange, particularly with discoveries or inventions supported by public monies.[64] Scientific discoveries are fostered by free exchange of ideas and information, and this value may also count as a return to taxpayers.

Yet another possibility exists for interpreting the taxpayers' fair share. One could argue that the return to U.S. taxpayers for support of basic research might come in the form of price controls on any medications or medical interventions that do result from the genome research.[65] In 1990, G. Kirk Raab, CEO of Genentech, testified to the importance of public support for research: "[Genentech has] discovered and developed four of the six biotechnology products that are on the market in the United States today. We have been able to do that because of the funding of the United States Government to the NIH and the great research and developments that have come out of that funding."[66] Even remembering that Raab is testifying in order to encourage federal support for the genome, it is significant to note that the Industrial Biotechnology Association that he represented at those hearings took cognizance of the role of public funding in supporting the biotechnology industry.

Raab is not alone in seeing how crucial public monies have been for the development of private industry. In the mid-1980s, 50 percent of research in the United States was conducted at universities, and 64 percent of that research was supported by public funds.[67] Moreover, Kenney notes that universities provide "free" trained labor to industry, and thus, one of the major costs of reproducing the entire economic system is already "socialized" or heavily publicly funded.[68] When a technology institute was established at Cornell University, it was originally intended to be supported by corporate sponsors. But the largest investments at the outset came from the New York state government and from the university itself rather than from industry. Nonetheless, industry benefits from the basic research done at the institute.[69] There is evidence that public monies already support private industry in numerous ways, and that indeed private biotechnology industry in the United States is heavily publicly subsidized.

In light of this public subsidy for private industry in the biotechnology arena, it is not outrageous to suggest that price controls on pharmaceuticals might be a fair "return" on the taxpayers' investment. Yet Raab is quoted as having warned the Clinton administration not to impose price controls on pharmaceuticals on grounds that it would cripple the nascent biotechnology industry by making it difficult for companies to raise needed capital.[70] What Raab appears ironically to neglect here is the very point that he made in his earlier testimony—that the U.S. public already supports the work of the biotechnology industry and therefore is entitled to a fair share of any return on that investment. The nature of that return should not be presumed to lie in the betterment of industry or in leaving its prices uncontrolled. As Negin observes, when "a corporation with exclusive rights to a patent produced by publicly funded research can sell the product at a monopoly price," the U.S. public has paid twice over for that research.[71]

Raab also ignores the extent to which that industry is dependent upon and linked with foreign capital. To get the capital for research and scale-up, most venture capital startup companies sell their services to other companies. They often sell to multinational pharmaceutical or chemical manufacturers.[72] For example, Genentech contracted with Mitsubishi Chemical and Kyowa Hakko of Japan, giving them marketing rights in Japan in return for research funding.[73] U.S. industry not only *competes* with Japan, but also *contracts* with and *depends financially* on the very Japanese business ability that is considered a threat to U.S. industry. Negin quotes the House Committee on Government Operations as reporting in 1992 that "the benefits of publicly funded research are being sold at bargain basement prices to foreign corporations."[74]

Given these international links, it is not clear that the best "return" for U.S. taxpayers is securing the first jump for U.S. biotechnology industry rather than imposing regulations or controls on pricing or retaining a system of open information and sharing. There are many other ways to interpret fair shares. Negin points out, for example, that patent laws give corporations exclusive licenses, but do not require that a corporation "work" a license—that is, develop it into a pharmaceutical or other intervention for use.[75] There is no guarantee that the public will ever receive returns on some research that it has supported.

The question of what constitutes a fair share for the public or for taxpayers is a question of who counts, what risks the public should be willing to take with its investments, and what would constitute the nature of the appropriate return on risk taking. One view predominated in the hearings, but other approaches can be defended. Secrecy in order to protect U.S. business interests is not necessarily the best way to ensure a fair share for taxpayers.

An Alternative Theory of Justice

All of these arguments have depended to a certain extent on the Lockean view of property rights, ownership, and proprietary interests. This is not surprising, since this view has been extremely influential in government circles, particularly during the Reagan and Bush years. In a manner similar to that of David Peters, who elsewhere in this volume distinguishes between a libertarian and an egalitarian matrix for pursuing justice, I want now to show the inadequacy of U.S. government thinking to date and to develop an alternative concept of fairness.

The Lockean view is not the only way to understand justice or what it implies for fair shares. Indeed, the very complexity of the issues around the genome raises questions about whether the Lockean view is adequate. As we have already seen, Japan may have done more work toward the genome than was credited to it. Suppose, for example, the little red hen's friends had tilled the soil and prepared the ground before she planted the seeds. Would they still have no claim on the final product? Or consider the process of commercial application. It is not as simple as eating the finished bread; it may take considerable effort, somewhat analogous to cutting, threshing, grinding, and baking the wheat. Doing the basic research on the genome may be more analogous to planting seeds and tending them until they grow than it is to the entire process undertaken by the hen. Separating out stages of research and calling some "basic" and others "applied" is not easy. Thus, we would be mistaken if we assume that the United States mirrors the actions of the little red hen, and Japan or other nations fit into the category of her friends.

Remembering, too, that much of the work on genetics uses data gathered from around the world, we might suggest that the original grains of wheat in the HGP are not simply "found" but are "borrowed." They do not begin as simple unadorned "nature," but as the product themselves of much collaborative work and effort. They often depend on the cooperation of peoples from many lands. Nancy Wexler's groundbreaking research on Huntington's chorea, for example, depended upon the participation of an entire village in Venezuela.[76] By agreeing to share their family histories, in essence the Venezuelans provided the initial grains of wheat from which the gene hunt could progress. It therefore might seem only fair that they would have a share in the end result. If diagnostic tests are developed in the United States under patent, however, the likelihood that those tests would be available—or affordable—to peasant villagers in the so-called Third World is slim indeed.

So it is questionable in the first place whether the Lockean approach to proprietary rights reflected in the story of the little red hen is adequate to a phenomenon as complicated as the Human Genome Project. Perhaps not everything that serves the common good should be able to be privatized simply by putting some effort into it.

Another story that many Americans hear growing up reflects a different view of fair shares. This is the story of the man with two sons.[77] The younger one asks for his inheritance while his father is still living. In his culture, it is a shocking thing to do; it is tantamount to treating his father as though he has already died.[78] But instead of rising up against the son in anger, the father gives him his inheritance. The son goes far away and there squanders his fortune. He is reduced to a point of destitution. So severe is his plight that he determines to return to his father and ask to be treated as a servant. That way, at least he will eat. But when he returns to his father's village in disgrace, his father runs to him, reinstates him as son, and calls for a public celebration. The son gets not the punishment that he deserves, but a gift of unmerited love.

However, the story does not end there. It is a story in three acts and the punch is in the third act.

There is another son, an older son, who is working in the fields. He refuses to join the celebration for his brother, saying to his father: "All these years I've been slaving for you and never disobeyed your orders. Yet you never gave me even a young goat so I could celebrate with my friends. But when this son of yours who has squandered your property with prostitutes comes home, you kill the fatted calf for him!" (Luke 15:29–30 NIV).

The older brother raises here an issue of fair shares: it is not fair that I, who have worked so hard all these years, get no rewards when the one who has not been responsible gets the fatted calf![79] His is the voice of protest of the one who sees himself as working hard only to have the fruits of his labor go to another. His is the voice of protest that might have been the little red hen's if her loaf of bread had been given to her lazy friends.

The perspective of the little red hen and that of the older brother are similar: only those who have worked hard should share the benefits; only those who have been there slaving from the beginning should get the rewards. The lazy friends, the profligate brother, deserve no bread, no fatted calves. It is not fair for them to share in the riches gained through the labor of others.

"In a limited goods society, receiving the younger son back must be at the detriment of the elder," says Scott in his interpretation of the parable.[80] The younger brother's good fortune does jeopardize the older brother's position. It does not seem fair. Those who have worked hard should get rewards for their labor.[81] Quite frankly, I have always thought the older brother had a point!

Yet another thematic is at work here. The parable challenges its hearers to move beyond the perspective of the older brother or the little red hen. The original audience, argues Scott, would have identified with the younger son.[82] In accord with the basic mytheme of younger and older brothers, the audience is primed to identify with the younger, who may be a rascal but is loved by his father, and to see the older brother as characteristically self-righteous and selfish.[83] Dom Helder Camara reflects on this parable in a sensitive poem, wherein he prays for the conversion of the prodigal's elder brother. Awakening from sin is one thing. Awakening from virtue is quite another.[84]

Those who identify with the younger son will see the "virtuous" older son who has labored so hard and thinks he is not receiving his fair share as being just as much in need of repentance as the profligate one. In fact, they expect the older brother to be rejected and denounced.

But that is not what happens.[85] Just as the father gave the younger son more than he deserved, so the father gives the older son more than he deserves. The older son has also insulted his father by refusing to share in the celebration. Yet once again, the father responds not in wrath, but in invitation: everything I have is yours and you are co-owner with me. In short, *both* sons have wronged their father, and *both* receive unmerited love. The father's concern is the unity of the family; for this, he has been willing to ignore the insults heaped upon him by both sons. If anyone has been treated unfairly, it is the father. Yet he responds with generosity and sharing.

In this parable, fair shares are determined not by looking to see who has mixed labor with the land, but by asking what serves the unity of the family. Both the self-righteousness of the older brother and the favored-child position of the younger brother are radically challenged by the parable. The dominion is universal, and everyone inherits all.[86]

If everyone inherits all, then everyone shares the "bread" at the end of the journey—not only those who planted and watered and threshed and reaped, but also those who made mistakes and bad judgments and squandered resources, and even those who were priggish and self-righteous. The glory is for all.

Fair Shares and Creative Tensions

The 1989 and 1990 hearings on the Human Genome Project incorporate a number of tensions around issues of fair shares. On the one hand, there is the desire to see U.S. business get a jump on international competition. This requires a structure that allows for secrecy and patenting. On the other hand, there is an ideal of scientific information as something that should be freely shared and available to all, not owned by some.

These contrasting views suggest underlying tensions in our understanding of fair shares. On the one hand, it seems fair that those who support research should reap the rewards deriving from it. On the other hand, it hardly seems fair for any nation to claim proprietary rights over something basic in nature and shared throughout the world.

At root is the question of what view of justice will inform our approach to this collaborative endeavor. We have basic concepts and values on a collision course. We cannot simultaneously hold that the genome belongs to no one and yet attempt to patent every piece of it. We cannot simultaneously hold that we value free exchange of ideas and yet support increasing structural changes that foster secrecy and undermine the free exchange of ideas. The Human Genome Project makes clear that in the arena of fair shares, we are trying to hold in creative tension differing notions of fairness. It remains to be seen whether they can be held together or whether they will ultimately shatter into fragments too fragile to mend.

Notes

1. The three hearings reviewed here are (1) U.S. House of Representatives, Hearing before the Subcommittee on International Scientific Cooperation of the Committee on Science, Space and Technology, 101st Congress, October 19, 1989 (hereinafter: House 1989); (2) U.S. Senate, Hearing before the Subcommittee on Science, Technology and Space of the Committee on Commerce, Science and Transportation, 101st Congress, November 9, 1989 (hereinafter: Senate 1989); and (3) U.S. Senate, Hearing before the Subcommittee on Energy Research and Development of the Committee on Energy and Natural Resources, 101st Congress, July 11, 1990 (hereinafter: Senate 1990). The purpose of these hearings was to determine whether Congress should support the Human Genome Project and appropriate substantial funds to undergird the mapping and sequencing efforts that had already been initiated by the Department of Energy and the National Institutes of Health.

2. Senate 1989, 84.

3. Another cluster focused on issues for scientists as a professional group and as individual practitioners. These issues will be reserved for another time.

4. Having served on a national commission that held many hearings during its tenure, and having testified before the state government of California, I am keenly aware of the limits of testimony at government hearings. Arguments made do not necessarily represent the personal priorities of those who testify, but may represent what they believe will be convincing to lawyers and policy makers, who are often responsive to cost-benefit analyses.

5. House 1989, 1.

6. Senator Hollings, Senate 1989, 5.

7. Senator Bryan, ibid., 55.

8. House 1989, 10.

9. Hall, ibid., 55.

10. Senator Pressler, Senate 1989, 3. In 1984, for example, the European Patent Office notified Biogen that it would receive the European patent for alpha-interferon; this would give it the exclusive commercial rights to the product in eleven European countries. See *Science* 223 (March 9, 1984): 1019. An editorial by Philip H. Abelson on commercial biotechnology suggested that the 1984 market value of products from nineteen new biotechnology firms was more than $2.4 billion. See *Science* 223 (March 23, 1984): 1253.

11. James Watson, House 1989, 65.

12. He estimated that at least 10 percent would go to private industry at the beginning of the project and more toward the end. Ibid., 58. As early as 1983, President Reagan sought to change patent policy so that all businesses, including large corporations, could retain patent rights on inventions made in the course of government-funded research and development work. See *Science* 219 (March 25, 1993): 1408.

13. Watson, House 1989, 58.

14. Leroy E. Hood, professor of biology at CalTech, Senate 1990, 91.

15. On June 18, 1993, the business section of the *San Francisco Chronicle* (D1) carried the headline "U.S. Trade Gap Widens" and began, "A widening in the trade gap with Japan . . ."

16. Hood, Senate 1990, 91.

17. Watson, Senate 1989, 52.

18. Hon. Ralph M. Hall, House 1989, 53.

19. Senate 1989, 12.

20. Domenici, ibid., 13.

21. Watson, ibid., 20.

22. Ibid., 52. Watson is somewhat known for his intemperate statements and should not be understood to hold particular animosity toward the Japanese government or people, and certainly not toward their scientists.

23. Wyngaarden, House 1989, 81.

24. Watson, ibid., 64.

25. Ibid., 61.

26. On the other hand, James Watson noted that "the vast majority of the basic information will be generated with public funds and be freely available." Ibid., 65. Thus, he appeared to distinguish levels of information sharing based partly on which funds supported the research.

27. Olson, Senate 1989, 115.

28. Ibid.

29. The practice of university-corporate contract has also raised numerous questions about the changing shape of the practice of science. These questions will be reserved for another discussion, though they will inform the critique below.

30. Elizabeth H. Wolgast, *The Grammar of Justice* (Ithaca, N.Y.: Cornell University Press, 1987), chap. 9.

31. Margot Zemach, *The Little Red Hen: An Old Story* (New York: Farrar, Straus and Giroux, 1983).

32. John Locke, "An Essay Concerning the True Origin, Extent and End of Civil Government," chap. 5, "Of Property." This is the second of Locke's *Two Treatises of Government*, published in 1690.

33. Senate 1989, 20.

34. Locke, "An Essay," chap. 5, par. 33.

35. Ibid., par. 36.

36. Robert Nozick, *Anarchy, State, and Utopia* (New York: Basic Books, 1974), 179.

37. Ibid., 180.

38. Ibid., 181.

39. Ibid., 182.

40. Marjorie Sun, "Strains in U.S.-Japan Exchanges," *Science* 237 (July 31, 1987): 478.

41. Ibid., 476.

42. Charles DeLisi, letter commemorating the retirement of Akiyoshi Wada, Boston University, January 2, 1990, quoted in Robert Mullan Cook-Deegan, *The Gene Wars: Science, Politics, and the Human Genome* (New York: W. W. Norton, 1994), 216.

43. Cook-Deegan, 218.

44. Ibid., 219.

45. Martin Kenney, *Biotechnology: The University-Industrial Complex* (New Haven, Conn.: Yale University Press, 1986), 244.

46. Quoted in Sun, "Strains in U.S.-Japan Exchanges," 476.

47. Olson, Senate 1989, 92.

48. Ibid.

49. George F. Cahill, House 1989, 46.

50. Ibid., 40.

51. Ibid., 46.

52. Fawell, ibid., 56. Even so, his view was predicated on the assumption that "all nations ought to want to cooperate somewhat on an equal basis."

53. Nozick would argue that proprietary claims over the genome would not worsen anyone's baseline situation.

54. The NIH applied for patents to protect a series of some three thousand partial gene sequences. See Bernadine Healy, "Special Report: On Gene Patenting," *New England Journal of Medicine* 327, no. 9 (August 27, 1992): 665. This action sparked vitriolic comment at the international meeting of HUGO in Nice, France, in November 1992. A CEO of a biotechnology company who asked not to be identified indicated to this author that companies are now seeking patents on everything they do and find because they do not know what may ultimately prove valuable.

55. House 1989, 66.

56. Biotechnology is also capital-intensive, however; most new companies show losses for several years and do not provide many jobs for the communities that woo them. See Colin Norman, "America Dominates in Biotechnology," *Science* 223 (February 3, 1984): 463.

57. One thinks, for example, of Birute Galdikas's groundbreaking study of orangutans, where it took her twelve years to habituate one orangutan to her presence. See Virginia Morell, "Called 'Trimates,' Three Bold Women Shaped Their Field," *Science* 260 (April 16, 1993): 420–25.

58. Kenney, *Biotechnology*, 31.

59. Ibid., 89.

60. The National Resources Defense Council, quoted in ibid., 31.

61. Jonathan King, professor of biology at MIT, testifying in 1981, quoted in ibid., 108.

62. Thomas F. Lee, *The Human Genome Project: Cracking the Genetic Code of Life* (New York: Plenum Press, 1991), 83–86.

63. According to Sayre, the information was conveyed through two routes. It was included in a routine report circulated to other laboratories by Sir John Randall, head of Franklin's laboratory. Under an informal understanding between laboratories, Randall would not have expected that anyone at Cambridge was working on the structure of DNA and therefore did not make the report confidential. Max Perutz of the Cambridge lab showed it to Watson and Crick. It was also given to Watson and Crick by Maurice Wilkins, who acknowledged that under usual protocol he would have sought Franklin's permission, but failed to do so because of the strained working relations between them. What is important is that Franklin had identified the helical structure of DNA in the work that was thus conveyed. See Anne Sayre, *Rosalind Franklin and DNA* (New York: Norton, 1975), 151–52. The issue of proprietary interest may be complicated by the fact that Franklin had unsuccessfully requested the information from Pauling's laboratory that became central to Watson and Crick's discovery. See Lee, *The Human Genome Project*, 83.

64. Watson also later objected to the broad plans for patenting being proposed by Craig Venter and Bernadine Healy. See Leslie Roberts, "Why Watson Quit as Project Head," *Science* 256 (April 17, 1992): 301–2.

65. The health care plan proposed by the Clinton administration in September 1993 includes some mechanisms for controlling the costs of pharmaceuticals. *San Francisco Chronicle*, September 10, 1993, A5. Among the prohibitions of the Lockean proviso, Nozick includes the possibility that the proviso might exclude persons charging certain prices for some of their supply when they hold a near monopoly on something. See *Anarchy, State, and Utopia*, 179.

66. Senate 1990, 101.

67. Kenney, *Biotechnology*, 35.

68. Ibid., 29.

69. Ibid., 47.

70. *San Francisco Chronicle*, June 16, 1993, C1.

71. Elliot Negin, "Why College Tuitions Are So High," *Atlantic Monthly*, March 1993, 43. Negin fails to establish that the link between the university and industry is the cause of increased tuition, but he does offer helpful analysis of what this link means for the university system.

72. Kenney, *Biotechnology*, 158.

73. Ibid., 160.

74. Negin, "Why College Tuitions," 44.

75. Ibid. The "orphan drug act" of 1983 is supposed to encourage development of pharmaceuticals under less lucrative circumstances by simplifying the testing process. See *Science* 223 (March 2, 1984): 914.

76. Lois Wingerson, *Mapping Our Genes: The Genome Project and the Future of Medicine* (New York: Penguin, 1990), chap. 10.

77. This story is usually—and probably wrongly—called the parable of the prodigal son. The title skews the reading of the story and obscures some of its important messages for fair shares. I have adopted Bernard Brandon Scott's approach of calling it the parable of the man with two sons. See Bernard Brandon Scott, *Hear Then the Parable: A Commentary on the Parables of Jesus* (Minneapolis: Fortress Press, 1989).

78. See Kenneth Ewing Bailey, *Poet and Peasant: A Literary Cultural Approach to the Parables in Luke* (Grand Rapids, Mich.: Eerdmans, 1976), 165.

79. "For the elder son, the younger is a profiteer who is depraved, while he is a slave to the father and faithful, never breaking a commandment." Scott, *Hear Then*, 121.

80. Ibid., 120.

81. It is partly this that makes it seem unjust, for example, that Rosalind Franklin was not recognized for her work that contributed a key element to the Watson and Crick discovery of the structure of DNA. See Lee, *The Human Genome Project*, 86.

82. Scott, *Hear Then*, 103.

83. Ibid., 121.

84. Dom Helder Camara, *A Thousand Reasons for Living* (London: Darton, Long-man and Todd, 1981), 71.

85. Scott interprets the basic point of the parable to be a rejection of Israel's self-understanding as the chosen people of God.

86. Scott, *Hear Then*, 125.

PART TWO

ETHICAL REACTIONS

[5]

Determinism, Freedom, and Moral Failure

Philip Hefner

IS THE HUMAN DESIRE to alter our genes itself grounded in our genes? Has our genetic evolution brought the human race to a point where we now can create a culture with its accompanying science and technology that will in turn guide the future of evolution? Is our freedom today something that has been determined by our genetic past? If this is in fact the case, then more ominous questions arise. If we have been determined to be free, and if freedom is the condition that makes serious human sinning possible, then do we now find ourselves at risk of contaminating, if not derailing, our evolutionary future? Might the freedom we have inherited provide the opportunity for human self-development but also the opportunity for moral failure on a colossal scale?

These questions arise because new and wide vistas of human possibility are being opened up by the frontier of genetic research. Seldom has a single scientific-technological procedure prompted more philosophical excitement or posed more profound theological and moral questions than what goes under the name Human Genome Project (HGP). Glimpsing the depth and breadth of the project, we see (1) the science that underlies the project provides understandings of human life that are both significant and difficult to take the measure of;[1] (2) when that science is linked to sophisticated technology, it will enable interventions in the genome that are capable of genuinely revolutionary alterations of human life;[2] and (3) these interventions in turn raise the most probing moral questions.[3] These three broad dimensions are intimately interre-

lated—scientific understanding, technological intervention, and moral issues.
The interventions are so exciting for the medical and technical practitioners
and so enticing for the general public that a powerful pressure is created for
practice to forge ahead even when understanding and moral reflection have
scarcely recognized where the problems lie, let alone provided responses to
them. We can throw light on the issues raised by these background factors and
their interrelationships in the context of examining three themes that are peren-
nial in the history of reflection upon what it means to be a human being: freedom, determinism, and moral failure.[4]

Freedom, Determinism, and the Evolution of the Genes

The science of genetics and its various subfields, along with many other scientific disciplines, play an important role in understanding the human being and
its evolution. The large picture that is emerging presents a *biocultural evolutionary paradigm* for interpreting *Homo sapiens*. In this paradigm, humans appear as
two-natured creatures, the confluence of two major streams of evolutionary information, the genetic and the cultural. Genetic evolution has brought humans
to the point where we are capable of becoming significantly cultural animals and
also dependent upon culture for successful adaptation to our niche and survival
in it. Culture in this context refers to learned patterns of behavior, in contrast to
the physico-chemically and -genetically programmed behavior that marks most
other things in the world. Culture includes these learned behaviors and the sym-
bol systems in which human beings contextualize these behaviors, so as to interpret and justify them. Birds build nests on the basis of genetic programs,
whereas humans have to learn how to build houses, and we pass on that knowledge and skill to later generations through the processes of teaching and learning, with the modifications that each generation makes on what it receives.

Humans are therefore genuinely biocultural creatures. Our biology, chiefly
our genes, assumed the configuration that makes culture possible; and having
come to that point, it also made culture necessary for our survival. The physico-
chemico-biological evolutionary processes in which we emerged, processes that
ultimately go back to the origins of the universe (perhaps in the big bang), transpire within a deterministic framework. Through their understanding of the
deterministic facets of these processes, physicists and chemists can make reliable predictions about the present and future states. Here, however, is where
things become interesting: the viability of this deterministic system of evolution apparently depends upon the presence of indeterminism (stochastic factors) and even freedom. Freedom and determinism must be understood and
defined within this evolutionary context. They cannot be abstracted from it, as

they generally are. Both terms are human constructs that have been fashioned to articulate certain basic features of our experience as we have reflected upon our lifeways within the evolutionary process in which we emerged and carry out our lives.

In this biocultural evolutionary context, some operational definitions come to the fore. *Freedom* is an element of our becoming that is marked by the ability and the necessity for exploring our environment for the sake of discovering new and more adequate ways of living; this exploring is marked in humans by conscious deliberation, decision, taking of responsibility, and a certain autonomy. *Determinism* is an element of our becoming that always exists within a context that includes both causal past and causal present and that gives a clue to the future, the context in which the four basic characteristics of freedom transpire. Determinism is the basis of what we are—the physical, chemical, biological, and cultural basis. Freedom is the exploratory dimension that enables this determinism to survive and become what it can be. In a sense, freedom is part of what has been determined to be.

Determinism is the polar concept of freedom.[5] The *is* of the evolving trajectory is precisely what appears to be determined, whether that be the four basic forces that physicists believe may be the foundation of big bang cosmology, the laws of particle physics that evolved later, the principles of biological evolution, or the cultural constraints of society and tradition. The causal relationships that have produced us and that form the structures enabling our experience are what determinism is all about, and the true *oughts* of this determinism are what freedom seeks to actualize in more adequate ways.

Our genetic constitution, as I have said, bequeathed to us the basic structure of who we are, and that included the possibility of culture. To be creatures of culture, we were created as creatures who cannot survive without culture. All of this is the determinism in which we exist. By definition, however, culture is an exploratory reality, and its survival depends upon the richness and competence of its explorations. In other words, the determinism that makes us what we are depends for its existence upon the freedom that discovers and makes real the possibilities in which that determinism can become more than it has been in the past.

We confront here the fundamental challenge of human nature. The human being is a two-natured creature, as I have implied: a coalition or symbiosis of genes and culture. "Human genes have accomplished what no other genes succeeded in doing. They formed the biological basis for a superorganic culture, which proved to be the most powerful method of adaptation to the environment ever developed by a species."[6] Culture has elaborated the freedom that our genes make possible. Our genetic heritage and the environment that we are evolving with provide us with our determined possibilities and constraints. Al-

though environment receives less attention in this chapter, it is a coequal partner with genes and culture in shaping human beings, whether at the micro- or
macrolevel. From all of this, we can fashion a description of human destiny.
Our destiny consists of the future to which these genetic and environmental realities can be brought by our culture so as to actualize the most desirable existence for the whole.[7] When we understand this nexus of factors, then we have
placed the freedom/determinism theme in its proper evolutionary context.

This biocultural evolutionary sketch, itself partially derived from genetic research, provides a framework for interpreting the Human Genome Project itself. The science of genetics and the map of our genes that it will provide when
it is applied through sophisticated technological processes and apparatuses reveal to us our genetic determinism (as the term is defined above). Our freedom
responds to that revelation in ways that we think will carry our genetic heritage
into more desirable channels and forms. We believe that eliminating gene-
based diseases and defects is a more desirable form of existence for the human
genome. Some of us might think that preserving male fetuses and aborting females are also more desirable. Or that tall boys are more desirable than short
ones. This is a remarkable characteristic of human beings, that we do not accept
the nature with which we were born to be necessarily the most desirable, nor do
we hesitate to attempt transforming it. We approach even the laws of nature as
challenges that can be altered. Consequently, on occasion we change the course
of rivers, employ artificial selection in growing plants and raising livestock, and
alter our physical bodies through medication, surgery, and the like. The nature
we have been given collides with our humanly constructed conceptions of what
nature should be. This is the exploratory freedom that our genetic evolution
has bequeathed to us, the exercise of which is essential both to our humanity
and to our survival. The chief instruments of this freedom and reshaping of nature are in our culture.

Moral Fallibility in Genetic Perspective

I have just sketched the basis for asserting that the desire and ability to alter our
genes are very much grounded in our genes, particularly as they been instru-
mental in our emergence as creatures of culture. But there is more to the story.
Our nature as genetic and cultural has also bequeathed to us a set of fundamental tensions that render us morally vulnerable and fallible as we set about
exercising our cultural possibilities.

One set of tensions is grounded in the dissonance resulting from the fact that
our genetic information unleashes programs that our cultural information either cannot or will not assimilate within its programs for our lives. We carry

within our genes the history of the planet's experiments with living forms. Paul MacLean, the brain researcher, underscored this aspect of human nature in his concept of the "triune brain." He spoke of a most primitive level, the reptilian, on which evolution grafted the paleo-mammalian brain, on top of which the distinctively human neocortex has evolved. The task of the human level is to teach the reptile and the ancient mammal within us to live like humans. In the light of current research, MacLean's picture is at best of heuristic value, but it makes an important point, namely, that our behavior is conditioned by information that derives from programs more ancient than the human species; these ancient programs are part of us.[8]

In some cases, we cannot actualize these ancient programs. For example, we cannot experience the spontaneity that is rooted in the absence of self-awareness, as much as we would like to. The birds of the air and the lilies of the field glorify God simply by being what they are. Our worship of God requires self-critique and intentionality. In other cases, we recognize that we must refuse to obey the ancient programs, even if we wish to. There are responses of violence toward the outsider, of males toward females, for example, that we often feel but we must restrain and channel in new directions if we are to be genuinely human. Both sets of examples give rise to alienation within ourselves. Since both sets of programs are very much part of who we really are, we find ourselves set against ourselves. Being human involves negotiating this inner chasm of alienation.

Another set of tensions may be rooted in the conflict between our individual and our social nature. As individuals, we are all genetic competitors with one another. Donald Campbell, the experimental psychologist, has made much of this. As individuals, we are conditioned significantly by the philosophy of the "selfish gene." Yet as humans, we are called to a very intense sociality. Life as we know it would not be possible apart from the complex social existence that we have evolved. Next to humans, the social insects are the most intensely social of creatures. The sociality of the bees, however, rests on the fact that they are all clones, due to the conditions of their birth from the queen. As clones, they are in no genetic competition with one another. We, in contrast, are the products of sexual reproduction, resulting in genetic diversity and the competition that it spawns. Consequently, according to Campbell, our social nature is always trying to counter our biological nature, so that genuine altruism and cooperation will be possible toward those with whom we are not kin relatives.[9]

A third kind of tension results from the innate finitude and fallibility of both our biology and our culture. Our genes have given us a brain, but it is not infinitely quick or large, and it takes years for it to mature. Our culture must evolve much more quickly than the millions of years that constitute genetic evolution. Ralph Wendell Burhoe has described this situation:

Living systems simply are not fully preadapted to all future contingencies. It would seem that we can epitomize the program of life as the unending search for the right code without our ever fully reaching it. . . . If the failures and inadequacies of the codes of right behavior of any time and place are always with us, to that extent we are always wrong, bad, and evil. And since in evolutionary pictures of life this is the case, we may say that humans in this sense are inherently wrong, bad, and evil. One finds this parallel to religious doctrines of original sin.[10]

Our culture must respond not only to the environment in which it is set, but also to its conscious perceptions of that setting. It must determine by its own decision which of alternative perceptions is to be given greatest weight, just as it must determine and authenticate in itself what is to be authoritative for its processes of selection and action. At every moment, the system is itself aware that it knows too little, that its projections are based on inadequate data, that its stamina is less than it desires and needs. This is inherent in the central nervous system processes that are at the heart of distinctively human life.

The upshot of all this is that we are innately incapable of carrying out the actions that are both possible and essential for human life in ways that are fully adequate. Successful exploration and innovation are what human beings are all about, and we construct visions of that success. We are made by nature, however, too fallible to construct wholly adequate visions and too finite to attain even the successes we have imagined.

What I have just surveyed is the biocultural evolutionary foundation for the experience that our religious and theological traditions have expressed and interpreted in the symbols and doctrines of the Fall, original sin, and actual sin. The Western theological tradition has reached a consensus on these symbols that can be summarized thus: (a) Sin is an inherent factor of our self-awareness. (b) We participate in sin as a condition pertaining to our very origin as persons. (c) Sin seems to be inherited in some fashion. (d) Sin is associated with our freedom. (e) Sin is marked by a sense of guilt and estrangement, thus requiring the gift of grace. These elements will figure in my discussion of the biological materials.

What have become problematic in our tradition are the notions that we have fallen from some pristine perfection and that we are sinful because of the sinful acts of some ancient forebears named Adam and Eve. Eastern Christian traditions have also differed with the West on the issues of the Fall and original sin, even though they share a consensus on the gravity and enduring character of sin. Even the Western traditions have carried two notions of original sin: as the *first* sin, and as the *primordial* sin of our origin (*vitium originis*) or the sin that "comes with the territory" of being human.[11] The theological traditions in East and West have recognized that both freedom and determinism belong in the discussion of sin. The determinism is clear in Augustine's insistence that we are "not able not to sin (*non posse non peccare*)." Both he and the Eastern theologian Gregory of Nyssa were also clear in their views that human beings are intrinsically free, and that our sinfulness is deeply entwined with the vulnerabilities of that freedom.

Neither the inheriting of sin, as the sin of our origin, nor the present actuality of sin is challenged by the biocultural pictures. On the contrary, they help us to understand more fully the depth of the experience that the theological doctrines of original and actual sin interpret. They do challenge the notion, however, that we inherit sin because our primordial parents engaged in a disobedient act. Paul Tillich's interpretation is supported, that the Fall and original sin are not historical fact, but universally valid myth, that "before it is an act, sin is a state."[12]

Sin and Our Genes

The bioculturally enriched version of original and actual sin that I have just described is relevant to the HGP in two ways. First, it suggests how we should understand the relationship between our genetic composition and sin. Second, it helps to explain the risk to which we fallible creatures open ourselves when we engage in genetic intervention.

Genes and Sin

Researchers and others who are informed and sophisticated in their understanding of genetics recognize that although virtually everything about us humans is rooted in our genetic composition, most of our characteristics are influenced not by one single gene, but by several genes, often in complex interaction with one another. Since there are more than six thousand single-gene human genetic diseases, it is not surprising that much popular discussion is concerned about discovering *the gene* for the major defects that we suffer from, and also for certain behaviors, such as alcoholism. Personality traits and behaviors are influenced by many genes, each contributing in small ways.[13] Consequently, they are not simply present or absent in a person, but manifested to a greater or lesser degree on a continuum.

In light of this complex genetic basis for behaviors and traits, it is an error of significant proportions to speak of a single gene for a particular sinful behavior. If behaviors such as alcoholism and homosexuality are grounded in multiple and complex gene interactions, the same must be said for violence, lust, sexual promiscuity, and the like. It is absurd to say that we are determined by our genes to entertain a certain sinful thought, or to engage in this or that sinful behavior.

Rather, as the previous discussion suggests, we are created as finite and fallible creatures who are shaped by a subtle interplay of the determinism that grounds our being, but requires exploratory freedom for its expression and fulfillment. In Burhoe's words, we "simply are not fully preadapted to all future contingencies . . . we can epitomize the program of life as the unending search for the right code without our ever fully reaching it."[14] This is the matrix for our sin and its unavoidability. The innate dissonances between genes and culture, and between culture and individual selfish nature, also play into this matrix.

The biocultural dimension is relevant here. The pleasurableness of mood-altering drugs and substances and the strong desire for them are no doubt conditioned by our genetic composition. For most of human history, these characteristics were beneficial because the substances themselves were scarce and only infrequently enjoyed, perhaps at times of community celebrations, for example. Today, however, due to technological advances, alcohol and drugs are far more plentiful and relatively inexpensive, with the result that the ancient gene-influenced behaviors are destructive and life-threatening. The biocultural factors are so complex and time-bound that it is unhelpful and even perverse to speak of the "genes" for alcoholism and drug addiction, apart from the full, complex situation in which they appear.

The conclusion to be drawn, therefore, is that even though sin, like everything else about us, is surely rooted in our genes, it is real only in a rich context of biological and cultural factors, and always affected by the environment in which we live. Our expectations of the HGP ought to be commensurate with our more complex understandings. Genetic research will be helpful in understanding sin if we put it in proper perspective, but much more has to be taken into consideration. A fuller biocultural understanding may be very helpful. In the case of addictions to mood-altering substances, for example, it is probably wiser to engage in public education, therapy, and control of manufacture and distribution of the substances, rather than to engage in fantasies about a genetic diagnosis or quick cure. The cultural sources of sinful behavior and the cultural reinforcement of freely accepted individual discipline, rather than depression or fear at factors of genetic determinism, ought to be at the center of our attention. Unfortunately, our culture would probably prefer a genetic intervention for controlling guns in the streets, over the education, social reform, and discipline that provide more likely answers to the problem. These comments bring us to consider how our genetically conditioned "sinful selves" relate to the HGP and its possibilities.

Sin and genetic intervention. I suggested earlier that the significance of the human species within the biocultural evolutionary processes is to work, through our culture, for the most desirable future that our genetic and environmental context as a whole can enjoy. If this assessment is correct, it is clear that the HGP and the interventions that it opens up are another extraordinarily significant instance of *Homo sapiens* attempting to realize this destiny. The project is a lucid example of human culture, in the form of science, technology, and medical practice attempting to facilitate a new future of possibilities for the nature that is within us (which will impact the nature that surrounds us).

We do well to recognize how lucidly the HGP epitomizes our fundamental human destiny because it puts things into perspective. If we fear the project or if we cheer it, we should understand that it is not different in kind from what we have always been about. Consequently, the interpretation of determinism and freedom that I proposed earlier applies very directly to the HGP. Determinism is recognizable in the form of our genome—we can change it all, but we cannot trade it in for another model. Determinism is also visible in the stage that our knowledge and technology have reached—we cannot outstrip them to any significant degree. Freedom is recognizable in our exploratory attempts to stretch the genome in ways that have heretofore never existed. And our "stretching" efforts will be guided by our visions of what is the most desirable future—certainly for ourselves and, if our spirits can also be stretched, what is most desirable for the whole context in which we live. We recall in this context the long and comprehensive efforts we have made to intervene in the genome of other animals and plants. Oranges without seeds, corn that resists pests, cows that give milk containing a human enzyme, hogs that provide low-cholesterol meat—these are the products of genetic intervention by humans that actualize concrete futures we have envisioned for other citizens of the natural world in which we live, to serve our own interests.

What we have suggested in the earlier interpretations of moral failure, that is, in original and actual sin, also applies directly to our explorations of the human genome and its possibilities. In setting the fact of our moral failure against the background of an interpretation of human destiny, we make an essential point: the HGP is not a base, sinful, antihuman enterprise; rather, it is a quintessentially human adventure that is unavoidably vulnerable to the same reality of sin and moral failure marking all human efforts to actualize our destiny. Christians believe that destiny is given to us by God, in our creation and redemption in Christ. The message of the gospel centers on this creation and redemption, but it also includes, integral to the message of grace, the word of original and actual sin. We may not understand the whys and wherefores of sin, but we acknowledge their reality and the significance of grace in the context of that sin.

Our earlier analysis underscored three elements of original sin, defined not as the first sin, but as the sin that pertains to our origin: the dissonance between the prehuman information that we carry in our genome and the distinctively human; the dissonance between societal culture and individual "selfish" human nature; and the innate fallibility and vulnerability that mark our character as humans. All of these elements condition our visions of what the HGP can accomplish, the ways in which we conduct the project, and the interventions that we hope to carry out on the basis of knowledge and skill attained through the project. To sketch only the bare minimum impact of sin, we mention that (1) genetic interventions will be manipulated by social class interests (the affluent will surely benefit more than other classes, and indeed harm may be inflicted upon the poor and marginalized persons in our society); (2) certain groups will always hope to use the information provided by genome mapping for their own interests (insurance companies, employers, lovers); (3) inevitably, alterations will be made both in individuals and in the germ line that will prove to have been unwise; and (4) the technological facets of the genome project will threaten to define human life in ways that violate basic human dignity.

The mandate that we derive from this sober awareness of human destiny and human moral failure cannot be that the HGP is evil or forbidden, but that we must carry out the project in ways that take full cognizance of and responsibility for our fundamental nature as creatures of destiny, sin, and grace. To date, we have only a dim idea of what it means for us to carry out the project in such an awareness. Religious communities, and Christian churches in particular, stand under the clear imperative to provide guidance for our culture in the age of the genome. In every aspect of our lives, we are creatures of God-given destiny who must conduct our lives under the conditions of determinism, freedom, and moral failure. Christians believe that we are also creatures on whom grace has been bestowed, and we believe that grace makes a difference. Now is the time to think through and articulate what that difference is with respect to the Human Genome Project.

Notes

1. See the articles in *Science* 262 (October 1, 1993): 5130.

2. Michael Ruse, *Zygon: Journal of Religion and Science* 19 (September 1984): 297–316.

3. Ronald Cole-Turner, *The New Genesis: Theology and the Genetic Revolution* (Louisville: Westminster/John Knox Press, 1993), 71–75.

4. Much of what follows is taken from Philip Hefner, *The Human Factor* (Minneapolis: Fortress Press, 1993), chaps. 6–8.

5. Paul Tillich, *Systematic Theology*, vol. 2 (Chicago: University of Chicago Press, 1957), 29ff.

6. Theodosius Dobzhansky, *The Biological Basis of Human Freedom* (New York: Columbia University Press, 1956), 121.

 7. Hefner, *The Human Factor*, 118.

 8. Paul MacLean, "The Brain's Generation Gap: Some Human Implications," *Zygon* 8 (March 1973): 113–27.

 9. Donald T. Campbell, "On the Conflicts between Biological and Social Evolution and between Psychology and Moral Tradition," *Zygon* 11 (September 1976): 167–208.

10. Ralph Wendell Burhoe, *Toward a Scientific Theology* (Belfast: Christian Journals, 1981), 65.

11. Theodore Tappert, ed., "Epitome of the Formula of Concord," in *The Book of Concord* (Minneapolis: Fortress Press, 1959), 466.

12. Paul Tillich, *The Shaking of the Foundations* (New York: Charles Scribner's, 1948), 155.

13. Dobzhansky, *The Biological Basis of Human Freedom*, 121.

14. Burhoe, *Toward a Scientific Theology*, 65.

[6]

Genetics, Ethics, and Theology: The Ecumenical Discussion

Roger L. Shinn

FOR MANY CENTURIES HUMAN beings have related their genetic awareness to their ethical and ontological beliefs. Ancient mythologies, the scriptures of various religions, the high philosophies of Plato and Aristotle—all show evidence of convictions and prejudices related to the genetic understanding of their times. Past genetic ideas were a combination of superstition, ideology, and surmises resting on folklore and common experience. Following the scientific discoveries of Mendel (1822–84) and Darwin (1809–82), there came a burst of genetic and eugenic argumentation, often in the form of social Darwinism. Later generations can recognize in it a weird mix of science, pseudo-science, racism, imperialism, and conjecture.

Taking a warning from this history, the present generation has been quick to look for the human implications of the discovery of the double helix in the structure of DNA by Francis Crick and James Watson in 1953, followed by experiments demonstrating the possibilities of recombinant DNA or gene-splicing. Like all expansions of human power, this one raised ethical questions. And it raised further questions about freedom and determinism, the manipulation of life, and what it means to be human. Religious communities have taken an active part in the discussions.

Much of this chapter is a narrative account of activities of the World Council of Churches (WCC) and of the National Council of the Churches of Christ in the U.S.A. (NCC). I include also some account of the relation of the ecumenical Christian discussions to other events in the public arena. I do not include the several churches that have done their own work in this area, except for two prominent examples near the end. The sources for this narrative are, for the most part, matters of open record, publicly verifiable. But as a participant in several of the events, I can occasionally peek behind the scenes at the unrecorded dynamics of the process. Along with the narrative, I attempt to assess the most important issues and controversies that arise in the process.

All this means that my subjective interpretations color the story, especially when I report ideologically charged controversies. I aim to be accurate and fair, but other participants will report these events with different accents.

The discussions that I describe show some of the ethical excitement raised by the discovery of the double helix and the new genetics. Most of these discussions preceded the formal launching of the project to map the human genome. However, they raise the questions that the current project accentuates. I am here centering my attention on those issues. I am therefore slighting some frequent themes in the church discussions, for example, the repeated objections to use of genetic techniques in biological warfare, the concerns about the economic aspects of scientific research, and the patenting of new forms of life. As a test issue, I emphasize particularly the concern about the possible altering of human germ cells because this discussion presupposes success (at least, partial success) in mapping the genome and transplanting genes.

I describe, first, the work of the WCC; second, that of the NCC; third, examples of interesting fallout from these activities; and fourth, two more recent declarations from the United Methodist Church and the Roman Catholic Church.

The World Council of Churches

The background of the work of the WCC can be traced to the major World Conference on Christians in the Technical and Social Revolutions of Our Time, Geneva, 1966. Although the conference made considerable effort to include scientists, it quickly showed that religious leaders could talk more fluently about social revolutions than about technical revolutions. Some of the scientists urged, and the conference formally recommended, that the WCC give more attention to the ethical significance of science and technology. Subsequently, the subunit on Church and Society did so, developing a continuing program "Faith, Science and the Future." The chief organizer for the program was the indefatigable Paul Abrecht, WCC secretary for Church and Society, a man with a genius for winning the cooperation of capable, often famous people of many nationalities, professional skills, and ideologies.

The new turn quickly acquired unexpected relevance as societies around the world responded both to new technological achievements and to new evidence of ecological perils. Events forced the WCC, sometimes reluctantly, to relate a new set of issues to its traditional concern for social justice.

When Church and Society sponsored a meeting of scientists in Geneva in 1970, the issues of biological science, and specifically of genetics, came to the fore. In 1971, Church and Society recommended specific inquiries on genetics. The next year the Central Committee of the WCC, with apparent anxiety, accepted the recommendation. A series of events followed over several years.

Round 1

The first step was a consultation, sponsored by Church and Society in consultation with the Christian Medical Commission, in Zurich, June 25–29, 1973. The thirty-five participants represented many disciplines: biology, genetics, medical practice including psychiatry, social work, law, politics, ethics, and theology. They came from the continents of Europe, North America, Africa, Asia, and Australia. The moderator was Charles Birch, the eminent Australian biologist and author who later won the Templeton Prize in religion. Among the group were R.G. Edwards, the British pioneer of in vitro fertilization and implantation; Bentley Glass, a former president of the American Association for the Advancement of Science; and Alexander Capron, who later became executive director of the President's Commission for the Study of Ethical Problems in Medicine and Biomedical and Behavioral Research (USA). The theologians came from many traditions, including Roman Catholic and Orthodox.

The consultation worked with nineteen papers written by participants and later published in the book *Genetics and the Quality of Life*. Whereas the chapters express the varied views of the writers, the concluding "Findings and Recommendations" represents a consensus coming out of the meeting. In this exploratory venture the panel intentionally raised more issues than it settled. It found possible conflicts between individuals and the interests of society (e.g., the welfare of an individual and the modification of the gene pool), between free choices and social compulsion. It questioned costly treatments available only to the privileged in a world where many people lack the most elemental health care.

The group took some stands that, in view of subsequent public controversies, appear bold. On genetic therapy, it said, "In principle there is no ethical difference between treatment at the level of the gene and treatment at the level of symptoms as in ordinary medicine."[1] Thus, it regarded the prospective replacement of a defective gene, via a carrier virus, as ethically comparable to an inoculation. It found no objection to the use of fetal tissue after the death of the fetus,[2] but could not agree on the ethics of experimentation on a living fetus or on the culture of embryos in vitro (for a very few days) in order to conduct experiments before destruction of the embryos. In the absence of agreement, it urged further conversations on the ethical validity of artificial insemination by donor, in vitro fertilization and implantation, and the rights of embryos and fetuses.

What accounts for the peculiar mix of boldness and caution in these findings? Partly the particular persons participating. Edwards was engaged in experiments that were, at that time, unprecedented and daring; in conversation he was persuasive and ethically serious, and the group had no desire to condemn his work. Others came from churches (particularly Orthodox and Roman Catholic) with very specific beliefs about the relation between sexual union and procreation, and about abortion. Nobody wanted to override their convictions.

The group endorsed genetic counseling with the primary aim of helping patients or clients. It emphasized the need for professional competence of counselors but insisted that sensitivity to the beliefs and consciences of counselees was as important as technical skill. It showed distrust of elites, insisting that ordinary people must participate in policy decisions rather than turn them over to experts. Granting that "democratic procedures" are "very imperfect," it insisted on a responsibility of government for establishing social policies and financing both research and care of patients.[3]

The consultation spoke only on its own authority, with no endorsement from the WCC. Its findings were widely distributed and discussed by councils of churches and ecumenical groups in many areas of the world. They were further discussed by one section of the WCC Assembly in Nairobi, December 1975. The assembly recommended continuing attention to the "ethical dilemmas arising from the application of modern biology to human problems."[4]

Round 2

At the World Conference on Faith, Science and the Future at the Massachusetts Institute of Technology (MIT), 1979, one major plenary session had the topic "The Biological Revolution: The Ethical and Social Issues." The two main speakers, in an unusual arrangement for the conference, were from the United States, but both sounded warnings that the developing countries wanted to hear. Jonathan King, a biologist at MIT, spoke from the political Left and expressed concerns about the influence of corporations and profits on research and development. (King later became one of the founders of the Boston-based Council for Responsible Genetics.) Karen Lebacqz, identifying herself with feminist and liberation theologies, urged a shift from questions of the "right" decision to issues of who has the power to make decisions. She saw in biomedical technologies a threat to "the meaning structure of our world" and asked for "new approaches to the source of ethical insight."[5]

Most delegates, more comfortable discussing social than technical issues, welcomed the two addresses. Lebacqz, however, evoked a protest from some of the Orthodox, who took her ethic to be too "situational."

Two scheduled respondents sparked other controversies. Gabriel Nahas, then of Paris but later a professor at the Columbia University College of Physicians and Surgeons, made a strong attack on "genetic meddling." He saw risks to "the evolutionary wisdom of millions of years," risks with irreversible results that could not be known to the generation that perpetrated them. In manipulating DNA, he said, "science has transgressed a barrier that should have remained inviolate."[6] Traute Schroeder, a geneticist on the faculty of Heidelberg University in Germany, objected to definitions of health that permit rejecting persons on the basis of minor anomalies. She pointed to "the danger of destruction" in the programs of "prenatal diagnosis and selective abortion, and also legal abortion for social reasons."[7]

One of the ten sessions of the conference had the topic "Ethical Issues in the Biological Manipulation of Life," and almost half its report dealt with genetic issues. The report begins with a sharp assault on "the tragic consequences suffered by many people as a result of past eugenic theories and practices."[8] It refers to the justification of slavery, the sterilization of underprivileged groups, and the Nazi racial atrocities. Most published ethical discussions of genetics pay attention to this sorry record, but usually, it is not the entry point for the discussion. Why was MIT different? The answer is the expansive international character of the conference and an ideological conflict that permeated many of the debates, whatever the particular subject. A look at that conflict will show why worldwide discussion of genetics is often quite different from discussion in North America.

Two clashing concepts of science emerged early and continued throughout the conference: science as a search for truth and science as an instrument of power. On the first day Robert Hanbury Brown, a British-Australian astronomer, spoke on the nature of science. Warning that science is a social activity that has been "industrialized" and has "allied itself with power," he nevertheless maintained that it investigates reality and produces "progressively truer images of the world."[9]

In a prepared response, Rubem Alves, Brazilian theologian and social philosopher, characterized science as an expression of the will to power. Choosing the title "On the Eating Habits of Science," he wittily likened scientists to wolves preying on lambs. The ideology of science, he said, "assumes the superiority of the western civilization," accuses other civilizations of superstition, and ignores its own superstitions.[10] The second prepared respondent, the Kenyan biologist W. Mutu Maathai, found science a testimony to "the God-like" in humankind. Although she pointed to exploitation and environmental destruction, she saw in science a contribution to "faith and hope."[11] Many agreed, but Alves struck the sparks in the conference.

In the following days many speakers took account of science as power. Even so, near the end of the conference a caucus of delegates from developing countries protested that their concerns had been neglected. Their designated speaker, Alves, reaffirmed his earlier message, this time changing his animal images to tigers and deer.[12] The group presented a statement to the conference. The discussion was confused because the statement, as distributed to delegates, was labeled a "Manifesto" and began: "We denounce science and technology." Several delegates attacked that opening line. It turned out that the distribution, through a clerical error, was a preliminary draft that had been revised to read: "We denounce the historical and current use of science and technology by industrialized and technically advanced societies, to serve military and economic interests which have brought about great sufferings to the people of the Third World."[13] Practically everybody agreed with that, but the earlier draft seemed, to many delegates, to reveal a real animus within some groups at the conference.

Hence the discussion of the evils of eugenics in Section IV expressed an ethos—not the only ethos but a prominent one—of the conference. Delegates came to realize that any talk of genetics brings to some people an immediate association with snobbery, racism, and oppression. Geneticists in Section IV pointed out that modern genetic science has been one powerful force in refuting racist fallacies, but that did not annul all suspicions. The outcome was that the section report, like many ecumenical documents, showed many signs of compromise. Perhaps wisely, it included more questions than assertions. It did not totally reject prenatal diagnosis and abortion of defective fetuses, but laid a heavy burden of proof on the processes. It questioned the criteria of health and defects. Without flatly opposing artificial insemination by donor and in vitro fertilization, it again raised warnings. It denounced sperm banks maintained for genetic reasons, and it pointed out that programs of "positive eugenics" have "caused traumas and tragedy for people of certain ethnic identities."[14] The report did not totally oppose genetic therapies that would alter human germ plasm, but raised grave questions about the possibility and preferred a shift of attention to "creating a society in which our genes are protected from unnecessary damage [by mutagens and radiation] and full scope is provided for the development of our existing capabilities."[15] It insisted on lay participation in policy decisions on genetic research and practice. And like all such discussions, it called for further studies.

Round 3

After MIT, Church and Society circulated the report of the conference, as well as its own interpretation of issues, to more than two hundred biological scientists for comments. Replies were both favorable and critical, with some scientists objecting that the documents exaggerated the risks of recombinant DNA technology. Church and Society, with the approval of the Central Committee of the WCC, then convened a working group investigating "Ethical and Social Issues in Genetic Engineering and the Ownership of Life Forms." The group met in Vogelenzang, Netherlands, June 15–19, 1981. The twenty members came from the United States, Eastern and Western Europe, India, and Australia. They included the usual mix of professional skills. Once again, Charles Birch chaired the group. And again Alexander Capron, who had then become executive director of the President's Commission in the United States, participated.

Like all WCC studies, this one gave considerable attention to the social context of genetic research and practice. It warned against past abuses of eugenics, commercialization of science, and exploitation of developing countries. More specifically than earlier studies, it concentrated on the issues of genetic therapy. It saw the alteration of genes in somatic cells as "comparable to other forms of experimental therapy"—that is, as justifiable provided very serious attention was given to risk and informed consent. On the more radical possibility of altering germ-line cells, if that should become feasible, it went into more detail than earlier WCC studies. It pointed out that such changes could have unknown and irrevocable consequences. Nevertheless, it said, "changes in genes that avoid the occurrence of disease are not necessarily made illicit merely because those changes also alter the genetic inheritance of future generations. By overcoming a deleterious gene in future beings, the beneficial effect of such changes may actually be magnified." Aware of objections to altering the germ line—the point stated forcefully by Gabriel Nahas at MIT—the group noted that mutation is common, with at least 10 percent of human zygotes containing "one mutant gene not present in either parent."[16] The implication was that scientists might legitimately learn to do what nature already does blindly.

The group gave attention to the conventional distinction between "negative" and "positive" genetic engineering, the former aiming to prevent obvious genetic "defects," and the latter aiming to "improve" the genetic constitution of persons or humankind. The ecumenical literature in general shows a recognition that judgments about sickness and health are culturally conditioned. But there are some liabilities that nobody would desire for self or offspring, whereas dreams of improving the human race show fantastic diversity of imagination and prejudice. The Vogelenzang discussions showed some difference of opinion, not specifically recorded in the report. Some persons advocated an absolute ethical, perhaps legal, prohibition against positive genetic activities. The majority disagreed, not out of enthusiasm for ambitious attempts to remake the human race, but on a narrower ground: "There is no absolute distinction between eliminating 'defects' and 'improving' heredity. Correction of mental deficiency can move imperceptibly into enhancement of intelligence, and remedies of severe physical disabilities into enhancement of prowess." Instead the report settled for a strong warning:

> For some parents, the absence of certain desired traits in their offspring may seem as great a burden as the presence of medically defined "defects" seems to others. It is the responsibility of the scientific community and other knowledgeable people to disabuse the public of the notion that such "positive" traits can at present be identified genetically, much less supplied at will in individual cases. More fundamentally—against the day when the capability for "positive" genetic engineering may be at hand—people should be encouraged to examine, in the light of ethical and religious beliefs, whether the use of such capabilities would be desirable. The transformation of procreation into a laboratory science and of offspring into products with interchangeable parts to be selected at will would bring about a major change in human self-understanding.[17]

Continuing this line of thought, the report saw in genetic manipulation a "tendency towards total reductionism," in which persons become objects to be manipulated.

> Many ethical and religious traditions endorse some human freedom to modify or transcend nature. But for us to change substantially the germ-line DNA is to directly alter the genetic foundations of the human. In what ways do we, by manipulating our genes in other than simple ways, change ourselves into something less than human? If we accept this power, do we have the wisdom to use it?[18]

The draft report was circulated to scientists for criticism, then edited and somewhat revised by the Working Committee on Church and Society. The Central Committee of the WCC in 1981 received the report "with appreciation" and, in a familiar phraseology, commended it to the member churches "for study." The Central Committee also asked that the recommendations concluding the report be transmitted to member churches, national academies of science, and specialized agencies of the United Nations. That action accorded the report a recognition, just short of formal endorsement, by the WCC. The report was published in 1982 as a thirty-six-page booklet with the title *Manipulating Life*.

Round 4

In 1983 the Vancouver Assembly of the WCC encouraged further attention to issues of "human values and ethics" in "biomedical technology" and "genetic engineering."[19] The next major work of the WCC is a report of Church and Society, *Biotechnology: Its Challenges to the Churches and the World*, published in August 1989. This thirty-three-page booklet was prepared by the staff of Church and Society and a five-person drafting committee over a period of several months. An initial draft was "sent to 100 experts for advice and comment." The document was revised in the light of criticisms, then submitted to the Central Committee, which made further revisions and authorized distribution "to encourage and facilitate discussion of these complex matters in the member churches."[20]

The document summarizes the work of the WCC since 1969 and carries forward many of the discussions of the past. Its most important feature is a single page of recommendations, approved by the Central Committee of the WCC. Some of them, important in themselves, are not directly related to the Human Genome Project (HGP): stern rejections of genetic testing for sex selection, of "commercialized child bearing" and "commercial sale of ova, embryos or fetal parts and sperm," of patenting of animal life-forms, and of genetic research directed toward biological warfare. There is a curious contradiction, perhaps deliberate, in the advice to governments "to prohibit embryo research, with any experiments, if agreed, only under well defined conditions"—apparently a more rigorous position than that of Zurich, 1973.[21]

Two of the recommendations are more specifically relevant to the mapping of the human genome. Here the WCC:

> b) Draws attention to ways in which knowledge of an individual's genetic make-up can be, and in some cases, is being abused by becoming the basis for unfair discrimination, for example, in work, health care, insurance and education. . . .
> d) Proposes a ban on experiments involving genetic engineering of the human germline at the present time, and encourages the ethical reflection necessary for developing future guidelines in this area; and urges strict control on experiments involving genetically engineered somatic cells, drawing attention to the potential misuse of both techniques as a means of discriminating against those held to be "defective."[22]

I should note, in light of later controversies, that the proposal would ban experiments on the germ line *at the present time*—a restraint that virtually all scientists concur on—and that the encouragement for development of "future guidelines" suggests that the proposed ban is not meant to be permanent. Thus, the proposal is quite consistent with the judgment of Vogelenzang, 1981.

Embedded in the total document (although not in the page of recommendations) is another issue: "But decisions regarding genetic characteristics are of the most intimate nature, and deeply affect our personhood. They must never be determined by commercial or political pressures."[23] The warning against commercialism runs through most of the WCC documents. But there is a profound ambivalence about political influences. The sentences just quoted are designed to protect persons against governmental intrusion into personal life and are quite understandable, given the history of totalitarian governments, ancient and modern. However, the recommendations appear to call for some governmental prohibitions (on genetic testing for sex selection, on modifying the human germ line at the present time, and on embryo research). And the report of Zurich, 1973, called for an active role of government and "democratic procedures" in relation to genetic experimentation and practice. Some sorting out of these issues remains to be done. The issue here is as old as civilization: the relation of personal freedom to society and governments.

After publication of this report, the Canberra Assembly of the WCC called upon churches to "study, reflect upon and address with sensitivity, ethical issues concerning bio-technology (including genetic engineering), prolonging of life, euthanasia, surrogate motherhood, etc."[24] We can expect further developments in coming years.

The National Council of Churches

Round 1

Following Round 1 of the WCC, the Governing Board of the NCC commissioned a task force to inquire into issues of "human life and the new genetics." The NCC designated Roger Shinn as chair and Karen Lebacqz as vice chair.

In the style of the WCC, the group included a variety of professional competencies and of gender, race, and social location—not simply because of an abstract desire for diversity or "political correctness," but because of the many ways in which genetic information and activity impinge upon different groups in society. As a task force of a national council, it did not include the international representation of the WCC. Among the several scientists was Bentley Glass, who had been active in the WCC, Zurich, 1973. Theological ethicists included John Fletcher, James Gustafson, and LeRoy Walters. Other members were counselors and administrators of clinics and research. A panel of consultants included such prominent specialists as Daniel Callahan, Alexander Capron, Theodosius Dobzhansky, Margaret Farley, and Margaret Mead. Joseph Fletcher and Paul Ramsey, about as far apart as two ethicists can get, were both helpful consultants.

The task force worked about three years, partly through holding occasional meetings and partly through mailing materials. It was commissioned to speak on its own authority and not formulate positions or policies for the NCC. That arrangement was mutually beneficial: it gave the task force freedom of method and judgment, and it spared the NCC the duty of passing on issues that it had not studied in detail. However, the Governing Board of the NCC received the report of the task force and commended it to the churches for study. The report was published, after some administrative delays and with only minor editorial tampering, as a fifty-seven-page booklet, *Human Life and the New Genetics.*

The ethical stance was generally similar to the prior work of the WCC at Zurich and the later work at Vogelenzang. There were the characteristic warnings against the intrusion of prejudice into genetics and eugenics. There were more specific warnings against the abuses of genetic screening. There were insistent reminders that society often turns genetic traits (e.g., skin color) into social liabilities. There was recognition that environment causes some genetic problems (e.g., the mutagenic effect of nuclear explosions and the depletion of ozone from the upper atmosphere).

On genetic therapy, the critical issue most relevant to the HGP, the report pointed to the benefits and risks of gene modification of somatic cells and the still graver issues that would arise if modification of germ-line cells should prove feasible. But it deliberately stopped short of asking for a prohibition of genetic interventions, negative or positive.

Compared with the earlier WCC document, this one is somewhat more pastoral in tone. It aims to help lay people understand issues and make decisions, rarely to tell them how to behave. (One exception: it regards abortion for reasons of the sex of the fetus—with the possible exception of sex-linked ailments—as "destructive of human dignity.")[25] In keeping with its pastoral character, it proposes roles of the churches in support, education, and advocacy of persons and families.

The document engages in direct theological discussion. It draws from theology not a set of prescriptions but "a context of awareness." On the familiar theme of "playing God," it affirms the Christian belief that God acts both through the impersonal processes of nature and through human activity. It shows political and ecological awareness that the expansion of human powers can be destructive. But it insists that human intervention in natural processes is not inherently wrong:

> Genetic practice . . . involves major interventions in extremely intricate processes that are not well understood. There is reason to ask of any specific genetic practice

whether or not it is a rash act with unpredictable consequences. There is reason to ask also, as in any human exercise of power, whether or not it is an instrument for a strong elite to impose its prejudices on the less powerful. Such questions may lead to an ethical judgment that some genetic possibilities should not be exploited, now or ever. But that judgment, if made, is a specific judgment, not a universal rejection of genetic activity. Theologically understood, God may work as truly through intentionally human genetic acts as through the humanly unintended genetic processes that have made humanity genetically what it is now.[26]

Similarly, the report deals with human dignity. It affirms that healing—now including genetic healing—can enhance human dignity. But it warns that "human dignity upsets the conventional equations of cost and benefit" and that the treatment of persons solely as objects of experimentation is an offense to human dignity. And it says, "If the success of genetics should convince us that only the normal can be loved and should thereby destroy compassion for the weak, the very successes would be a blow to human dignity."[27]

Round 2

Subsequently, the NCC instituted a Panel of Bioethical Concerns, which produced an eighty-one-page book, *Genetic Engineering: Social and Ethical Consequences*, in 1984. Chronologically, this followed Rounds 2 and 3 of the WCC's studies. It differs from the other documents (WCC or NCC) in two principal ways. First, it is designed quite specifically as a study guide for use in churches. Each chapter ends with "Questions to Consider" and "Suggestions for Study Leaders." The final eleven pages are a questionnaire, designed for return to the NCC. Second, it is an in-house production. The panel of twenty-five, chaired by Bishop James Crumley of the Lutheran Church in America, included ten members of the Governing Board (from ten member churches), six members of NCC units, five members of the NCC staff, and a writing team of three. Rena Yocum was the "principal writer." Frank M. Harron then incorporated the report into a book, adding other materials. Despite the usual disclaimer on the title page ("This is a report for study and comment, not an official policy statement of the NCC/USA"), the NCC has a higher degree of "ownership" than either the WCC or the NCC has in the other documents I have described.

The surprising feature is that the panel included no scientists or medical specialists. However, the editor explains, the panel and its members consulted scientists; and the book, with respect for scientific competence, reprints eleven pages of informational excerpts from *Splicing Life*, the government-sponsored book that will be described below.

I need not repeat here the many ways in which the book reinforces earlier WCC and NCC publications. More important, in view of later controversies and the HGP, is its argument on genetic therapy. After discussing proposals for such therapy, both on somatic cells and on germ cells, it says,

> Stopping any enterprise out of a fear of potential evil not only deprives humanity of the fruits of new findings but also stifles strong impulses for innovation and change. Nevertheless, the technological allure of gene splicing ought not to be allowed to blind society to the need for sober judgments, publicly arrived at, about whether there are instances in which the price of going ahead with an experiment or an innovation will be higher than that paid by stopping the work.[28]

Continuing its analysis, the book calls for restraint from experiments involving "inordinate" risks. It criticizes "the theory of eugenics" with its criteria of superiority and inferiority, and with cultural biases that "could militate against ethnic minorities, women, and, of course, all who are physically and mentally 'abnormal' or 'defective.'" Pointing out that a "positive eugenics program . . . would probably require social organizing and involuntary compliance," it finds "involuntary eugenic research morally unacceptable."[29]

That comes close to a rejection of "a positive eugenics program." Is it a rejection of all genetic therapy involving germ cells? Possibly, since such therapy is involuntary for all future persons involved. But not quite, since the consequent recommendation is "extreme caution." And the next paragraph states, "We celebrate the healing possibilities that are now before us through this new life-technology."[30] The next chapter, "Values, Decisions, and Public Policy," comes close to a definitive judgment: "The manipulation of genes in human germ or sex cells, if permitted at all, should be subject to special scrutiny, because not only the individual, but also his or her descendants, may be affected. In such cases, it may be appropriate to require the approval of a guardian ad litem appointed to represent the interests of future descendants."[31] That is not a categorical rejection.

Round 3

Two years later in 1986, the NCC issued a Policy Statement—a sixteen-page pamphlet, "Genetic Science for Human Benefit." Chronologically this, like Round 2, came between Rounds 3 and 4 of the WCC processes. As a Policy Statement, adopted by the Governing Board of the NCC, this has a formal standing unlike any of the other WCC or NCC documents. It is not simply commended to the churches for study; it represents the position of the NCC. It is written with scientific, ethical, and theological thoughtfulness.

On many issues it reaffirms early ecumenical documents. It warns against elitism and prejudice that may infect genetic attitudes and practices. It denounces biological warfare. It raises economic issues forcefully but less dogmatically than some of the other ecumenical statements. It asks whether competition between commercial corporations, with its accompanying protection of scientific secrets, corrupts the traditional ethos of science. It affirms the value of human persons against any "ideology of genetic behaviorism which implies that levels of human action and social worth are genetically determined."[32]

Important though these matters are, I shall here concentrate on the one issue most relevant to the HGP: human genetic therapy, especially on germ-line cells. Here it states: "Given the primitive state of this art, as well as the unknown and uncontrollable risks involved in possible genetic mutations, researchers and religious and secular commentators should approach it with extreme caution." Later, returning to the subject, it says: "Gene-therapy of germ-line, or sex, cells of human embryos—if ever practicable—will deserve especially stringent control."[33] Once again, consistent with the other ecumenical documents, there is a strong warning against rash acts, but there is no dogmatic prohibition of genetic interventions.

Some Curious Fallout

The churches have no illusions that they can or should be ethical arbiters for society, whether in the United States or in other countries. But they seek to contribute to the public conversation with the hope that their ethical and theological insights may evoke resonance in the general public. In the United States two public events are closely related to the activities of the NCC. Historical causation is rarely so precise as to allow direct statements that one event caused another. But clearly, there has been a relationship, in some ways ironic, between the NCC and public and governmental discussion.

The President's Commission

On June 20, 1980 (between Rounds 1 and 2 of the NCC process), the general secretaries of three religious organizations joined in a letter to President Jimmy Carter. The three were Dr. Claire Randall, NCC; Rabbi Bernard Mandelbaum, Synagogue Council of America; and Bishop Thomas Kelly, United States Catholic Council. They called attention to "a new era of fundamental danger triggered by the rapid growth of genetic engineering." Calling attention particularly to the issue of patents on new life-forms, which had been recently validated by the Supreme Court, they asked for attention to "the entire spectrum of issues involved in genetic engineering." And they requested the president to "provide a way for representatives of a broad spectrum of our society to consider these matters and advise the government on its necessary role."[34]

The president's science adviser then asked the President's Commission for the Study of Ethical Problems in Medicine and Biomedical and Behavioral Research, an already active body, to add this issue to its agenda. The executive director of the commission was Alexander Capron, a lawyer specializing in medical issues, already twice mentioned in this chapter because of his participation in WCC studies. The eleven-member commission, headed by Morris Abram, included one theologian, Rabbi Seymour Siegel of the Jewish Theological Seminary of America. The commission, in addition to its other work already in progress, produced a 115-page book, *Splicing Life: The Social and Ethical Issues of Genetic Engineering with Human Beings*. The book came out in time for its use in Round 2 of the NCC processes, as noted above.

Appropriately for a governmentally sponsored inquiry, the book is reasoned in terms of the general values of a pluralistic society, but includes a sensitivity to the values of particular religious traditions. It even takes up the theological question of "playing God." Its assessment of issues is remarkably similar to those of the group inquiries of the WCC and NCC, except that it says little about the international ideological issues important to the WCC. It tends to be less alarmist or more complacent—take your choice—than some sentences of the religious documents, including the letter to President Carter that sparked the process. Participating in its inquiries at one stage or another were many religious ethicists, including Seymour Siegel, J. Robert Nelson, LeRoy Walters, John Fletcher, James Childress, Charles Curran, Margaret Farley, James Gustafson, Karen Lebacqz, Donald G. McCarthy, Paul Ramsey, Harmon L. Smith, and Gabriel Vahanian.

On the issue of genetic therapy, the commission found the issues not "appreciably different from those involved in the development of any new diagnostic and therapeutic techniques"—except for the influence on future generations, which calls for "especially stringent" precautions. To those who objected, it replied (with a shrewdly placed footnote to Pope John Paul II): "To turn away from gene splicing, which may provide a means of curing hereditary diseases, would itself raise serious ethical problems."[35]

The commission came close to summing up its conclusions in this paragraph:

The Commission could find no ground for concluding that any current or planned forms of genetic engineering, whether using human or nonhuman material, are intrinsically wrong or irreligious per se. The Commission does not see in the rapid development of gene splicing the "fundamental danger" to world safety or to human values that concerned the leaders of the three religious organizations. Rather, the issue that deserves careful thought is: by what standards, and toward what objectives, should the great new powers of genetic engineering be guided?[36]

Yet the commission saw the need for "periodic reassessment" of issues "under the scrutiny of a free press and within the general framework of democratic institutions."[37] That clearly made room for the continuing inquiries of religious organizations.

Coincident with the publication from the commission, then Representative (now Vice President) Al Gore conducted hearings of the Subcommittee on Science, Research and Technology of the House Committee on Science and Technology. On November 17, 1982, he invited religious representatives, in the typical triadic pattern, to testify to the committee. The three were Seymour Siegel (Jewish), Richard McCormick (Roman Catholic), and Roger Shinn (Protestant). Other theologians, including John Fletcher, also testified, but not as part of the hearing specifically designated for the religions. Gore's questions showed him to be knowledgeable on the issues. After hearing from the three, he asked whether they were all recommending a green light for continued genetic research. One replied, "A yellow light: proceed with caution." The others assented. All three agreed that they saw no need for legislative restrictions at that time, but that a continued monitoring process would be desirable. Gore soon introduced a bill into the House of Representatives to establish a commission for continuing oversight. So far as I know, it was never enacted. But the Congress, in authorizing the mapping of the human genome, specifically included a provision requiring research on its social and ethical implications.

Jeremy Rifkin's Statement

In June 1983, about half a year after the book from the President's Commission, author and crusader Jeremy Rifkin released a resolution affirming that "efforts to engineer specific traits into the germline of the human species should not be attempted."[38] The accompanying press release began: "The nation's religious leaders will call upon Congress to prohibit genetic engineering of the human germline cells at a press conference." It also claimed endorsement of the resolution by a wide range of persons, including "virtually every major Protestant leader"—thus earning the resentment of many who did not sign but liked to think of themselves as leaders. Rifkin, to do him justice, gained endorsements from the chief officers of most major denominations and from a remarkable combination of "mainline" and conservative evangelical leaders, along with several Catholic bishops, one rabbi, and a few scientists.

The basic proposition was unqualified; it did not say "at this time" or "until adequate guidelines are determined." The fact was that nobody was proposing to modify human germ cells in any immediate future.

Alexander Capron promptly criticized the statement in an op-ed piece in the *Washington Post*.[39] He flatly denied its claim that "it will soon be possible to engineer and produce human beings by the same technological design principles as we now employ in our industrial processes." Any such possibility, he said, is remote.

The surprise was that so many religious officials signed the resolution. It is a statistical certainty that several of them, probably a majority, had voted in the WCC or NCC to accept the documents I have described above. James R. Crumley Jr., presiding bishop of the Lutheran Church in America, had chaired the NCC Panel of Bioethical Concerns described above in Round 2 of the NCC processes. Granted, the votes of acceptance did not mean endorsement of everything in the WCC and NCC documents; usually, it meant commendation of the statements as educational documents for the churches. But at a minimum, the religious leaders knew—if they had carefully read the statements they voted on—that Rifkin's position differed from those statements. Granted, no document directly endorsed modification of the human germ line, least of all "positive eugenics." And Rifkin's resolution could be interpreted, although his intention was far broader, as directed only against positive eugenics. The ecumenical documents generally edged up to the issue, then deliberately stopped short of direct condemnation of human genetic intervention—or even suggested circumstances in which modification of the germ line might become appropriate.

Rifkin apparently did not realize that. On September 2, 1984, as it happened, I joined Rifkin and the redoubtable Harvard scientist Bernard Davis in a panel on the issue at the annual meeting of the American Political Science Association in Washington. Rifkin specifically claimed the support of both WCC and NCC documents. When I challenged him on the basis of direct knowledge, he answered that he had been "told" that the church councils agreed with him.

He also made clear that his objection was not solely to "positive" genetic intervention, as might be implied in the language rejecting "efforts to engineer specific genetic traits into the germline of the human species." Davis specifically asked Rifkin whether he would oppose the elimination of a disastrous hereditary disease if that should someday prove possible without harmful side effects. Rifkin's answer was that he would do so rather than open the door to a whole process that would be harmful.

I report this not to advocate a leap into the dark in genetic exploration.

Rifkin's argument deserves a hearing in ongoing conversations. A few—very few—eminent scientists agree with him. I have pointed to the beliefs of Gabriel Nahas in the WCC deliberations. Such voices should not be silenced. But the story is instructive because it illustrates the problem of relating basic religious-ethical convictions to specific decisions in a world of advancing science and high technology. We can guess why so many church leaders signed on with Rifkin, despite their prior approval of contrary position papers of the WCC and NCC.

One reason is that some precise distinctions are easily lost in hurried debate. In the consideration of a distant possibility, perhaps forever an impossibility, there is a world of difference—in ethical theory and in practical policy—between declaring that the world is not ready for such acts and pronouncing the acts wrong in principle. That difference was important in the ecumenical studies; it got lost in Rifkin's resolution.

A second reason has to do with the pressures operating on decision makers in our world—whether in government, business, or the churches. Imagine the church executive, receiving piles of mail each day, moving from meeting to meeting, writing speeches and sermons, considering a budget and personnel policy and theology and travel schedules. A letter asks endorsement of a statement on genetic interventions. The official remembers a past meeting where, among many votes, he or she more or less endorsed a statement expressing ethical concern about genetics. Now a decision is required: to sign or miss the opportunity. One possibility would be to phone members of the church who helped formulate those past statements. But their names and phone numbers are not on the desk at the moment, and there are many phone calls to make that day. So, maybe the decision goes, it's better to sign than be timid.

A third reason is "the hermeneutic of suspicion" that pervades our culture, with good reason, even where the phrase is unknown. We have too often been betrayed by experts who assured us about a light at the end of a tunnel, about the magic of the market, about tax cuts that would increase revenues and reduce deficits, about waste disposal that would not pollute, about weapons so invincible as to guarantee peace, about ways to economic justice that would hurt nobody because economic growth would make everybody rich. So when we are urged to distrust experts on genetics, we may be ready to respond.

We have also been confronted by many a fait accompli because we did not wake up to a danger until it was upon us. So if somebody alerts us to a distant peril, we may reckon that the time to protest is now, not when the peril is imminent.

All this, I say, is a guess as to why some people enlisted in Rifkin's crusade. What I know is that some of the signers later regretted their act, explaining it in terms something like those I have just suggested.

Two More Recent Declarations

In emphasizing the ecumenical discussions within the WCC and NCC, I have made no attempt to survey the positions of the various church groups that have given attention to ethical and theological issues in the new genetics. However, I here make two exceptions, partly because I want to update the record and partly because these two actions have received wide public attention.

One is the 1992 edition of *The Book of Discipline of the United Methodist Church*, which includes three paragraphs on genetic technology. In the two preceding years a United Methodist Church Genetic Task Force had prepared a report that was accepted by the church.[40] The paragraphs in the *Discipline* are much briefer, but give the issue prominence in the formal ethical teaching of the church. They "welcome the use of genetic technology for meeting fundamental human needs for health, a safe environment, and an adequate food supply." They warn against political or military "abuse of these technologies," as well as "unanticipated harmful consequences."

The most crucial paragraph then states:

Human gene therapies that produce changes that cannot be passed to offspring (somatic therapy) should be limited to the alleviation of suffering caused by disease. Genetic therapies for eugenic choices or that produce wasted embryos are deplored. Genetic data of individuals and their families should be kept secret and held in strict confidence unless confidentiality is waived by the individual, or by his or her family, or unless the collection and use of genetic identification data is supported by an appropriate court order. Because its long-term effects are uncertain, we oppose genetic therapy that results in changes that can be passed to offspring (germ-line therapy).[41]

That appears to be a more unequivocal rejection of germ-line therapy than any in the documents of the WCC or NCC.

The second document is the new *Catechism of the Roman Catholic Church*, issued in Paris on November 16, 1992, by a commission headed by Cardinal Ratzinger. Interpreting the Fifth Commandment, "you shall not kill," the catechism gives the Catholic doctrine on respect for human life. Here it repeats traditional teachings on contraception, artificial insemination, and abortion. It accepts prenatal diagnosis intended to safeguard and heal the fetus, but not with the prospect of abortion. The teaching on genetic therapy is, except for a few words, a repetition from the Instruction, *Donum Vitae*, issued by the Congregation for the Doctrine of the Faith on February 22, 1987, as indicated by the quotation marks within the citation:

"One must hold as licit procedures carried out on the human embryo which respect the life and integrity of the embryo and do not involve disproportionate risks for it, but are directed toward its healing, the improvement of its condition of health, or its individual survival."

"It is immoral to produce human embryos intended for exploitation as disposable biological material."

"Certain attempts to influence chromosome or genetic inheritance are not therapeutic but are aimed at producing human beings selected according to sex or other predetermined qualities. These manipulations are contrary to the personal dignity of the human being and his integrity and identity" which are unique and unrepeatable.[42]

In all the discussions of the churches on the theology and ethics of genetics, there is an awareness that the present generation is feeling its way toward an ethic that meets issues never quite anticipated in traditional faith and morals. There is a conviction that new genetic capabilities may enhance or jeopardize human dignity and the unique worth of human personality. There is the double awareness that it is important to raise ethical flags before the swift pace of technology settles issues without adequate consideration; yet the churches do not want to rush into dogmatic judgments without full consideration of new information and new potentialities.

This will not be an easy time for theological ethics. It must avoid the a priori judgments that stem from either technological messianism or technological demonism. It must distinguish, in a very confusing world, between real and fancied opportunities and perils. It must investigate issues with a rigorous examination of scientific evidence and a profound probing of its own beliefs and commitments. It must break out of ideological cocoons enough to understand why opinions that look mistaken to some people and communities are persuasive to others. Then it must communicate convictions, often complex and nuanced, in a world that longs to believe that everything important can be said in fifteen-second sound bites.

As physiologist R. G. Edwards told the first WCC consultation on genetics, "Society is often unprepared for making decisions on new developments," and the first responses to new possibilities are often "haphazard."[43] Too many decisions are made inadvertently as new opportunities generate their own enthusiasm and momentum. We may hope that adequate discussion, involving both specialists and the general public, will precede the decisions that follow from the mapping of the human genome.

Notes

1. Charles Birch and Paul Abrecht, eds., *Genetics and the Quality of Life* (Elmsford, N.Y., and Potts Point, New South Wales, Australia: Pergamon Press, 1975), 209.

2. Ibid., 211.

3. Ibid., esp. 213, 218, 221.

4. *Breaking Barriers: Official Report of the Fifth Assembly of the World Council of Churches*, Nairobi, Kenya, 1975 (Geneva: WCC, 1976), 128.

5. *Faith and Science in an Unjust World*, Report of the World Council of Churches' Conference on Faith, Science and the Future, vol. 1, Plenary Presentations, ed. Roger L. Shinn (Geneva: WCC, 1980), 281.

6. Ibid., 281–82.

7. Ibid., 284.

8. *Faith and Science in an Unjust World*, Report of the World Council of Churches' Conference on Faith, Science and the Future, vol. 2, Section Reports, ed. Paul Abrecht (Geneva: WCC, 1980), 49.

9. *Faith and Science in an Unjust World*, 1:32.

10. Ibid., 43.

11. Ibid., 44.

12. Ibid., 373–76.

13. *Faith and Science in an Unjust World*, 2:171.

14. Ibid., 52.

15. Ibid., 54.

16. *Manipulating Life: Ethical Issues in Genetic Engineering* (Geneva: WCC, 1981), 6–7. See Roger L. Shinn, *The New Genetics* (Wakefield, R.I., and London: Moyer Bell, 1996), 137–38.

17. *Manipulating Life*, 7.

18. Ibid., 8.

19. *Gathered for Life: Official Report of the Sixth Assembly of the World Council of Churches*, Vancouver, Canada, ed. David Gill (Geneva: WCC, 1983), 69.

20. *Biotechnology: Its Challenges to the Churches and the World*, Report by World Council of Churches Subunit on Church and Society (Geneva: WCC, 1989), 1.

21. Ibid., 2.

22. Ibid.

23. Ibid., 7.

24. *Signs of the Spirit: Official Report of the Seventh Assembly of the World Council of Churches*, Canberra, Australia, ed. Michael Kinnamon (Geneva: WCC, 1991), 69.

25. *Human Life and the New Genetics: A Report of a Task Force Commissioned by the National Council of Churches in Christ in the U.S.A.* (New York: NCC, 1980), 45.

26. Ibid., 43.

27. Ibid., 44.

28. *Genetic Engineering: Social and Ethical Consequences,* prepared by the Panel of Bioethical Concerns, NCC, ed. Frank M. Harron (New York: Pilgrim Press, 1984), 18.

29. Ibid., 26.

30. Ibid., 30–31.

31. Ibid., 40.

32. "Genetic Science for Human Benefit," a Policy Statement of the National Council of Churches in the U.S.A. (New York: NCC, 1986), 6.

33. Ibid., 4, 13.

34. President's Commission for the Study of Ethical Problems in Medicine and Biomedical and Behavioral Research, *Splicing Life: The Social and Ethical Issues of Genetic Engineering with Human Beings* (Washington, D.C.: Government Printing Office, 1982), 95–96.

35. Ibid., 61, 56.

36. Ibid., 77.

37. Ibid., 78–79.

38. "Resolution," signed by Jeremy Rifkin et al. (Washington, D.C.: Foundation on Economic Trends, 1983).

39. Alexander Capron, *Washington Post,* June 16, 1983, A29.

40. Ronald Cole-Turner, *The New Genesis: Theology and the Genetic Revolution* (Louisville: Westminster/John Knox Press, 1993), 76.

41. *The Book of Discipline of the United Methodist Church,* 1992 (Nashville: United Methodist Publishing House, 1992), 97–98.

42. *Catechism of the Roman Catholic Church,* translation © by the United States Catholic Conference (Mahwah, N.J.: Paulist Press, 1994), par. 2275, 549.

43. R. G. Edwards, "Judging the Social Values of Scientific Advances," in *Genetics and the Quality of Life,* 40.

[7]

Genetics, Ethics, and Theology: The Roman Catholic Discussion

Thomas A. Shannon

THE HUMAN GENOME PROJECT (HGP) will provide one of the most comprehensive and fundamental perspectives on human nature and behavior. Although the map of the genes coming from the HGP will not resolve many of the classic issues in philosophy or theology, this map will be the basis for reconsidering both classical and contemporary perspectives on human nature and particularly human behavior. The impact of this cannot be underestimated. Additionally, this map will be the foundation for many medical or therapeutic interventions. The success of genetic diagnoses, genetic therapies, and even genetic engineering in many ways depends on the success of the HGP.

Whether the HGP succeeds or fails, our society in general and our religious leaders in particular will have to confront profound questions that can be dubbed only philosophical or theological. Many of the questions will be new and even many of the old ones will be significantly reframed by new genetic knowledge, but some frameworks and insights already exist with which to begin our evaluation of the ethical and theological implications of the HGP. This chapter presents various dimensions of Roman Catholic theology as a preliminary identification of several core issues. While I raise problems and issues from the perspective of only one Christian communion, several of the concerns raised are at least ecumenical in their significance, if not thematic issues for the community at large.

Foundations in Roman Catholic Thought

Natural Law

No one can understand the Roman Catholic mind without first understanding the basics of natural law. The initial framework within which Catholic theologians work to address contemporary questions is their classical training in natural law theory, a theory enunciated authoritatively by Thomas Aquinas in the twelfth century.

Thomas Aquinas developed a synthesis of two major traditions, the Stoic and the Aristotelian, which he combined with the common theology of the church, previously articulated in particular by Augustine. From Aristotle, Aquinas took the idea of society as an order existing for the mutual exchange of services for the common good. This suggests that society, and implicitly government, is a part of the structure of nature. Within this context, law functions not simply as a regulatory agency, but as a part of the universal structure of creation that derives its rational character from the intelligent plan of the Creator. Aquinas accepted from Aristotle the concept of a hierarchically ordered universe with definite structures ordained by the Creator.

From the Stoics, Aquinas borrowed the concept of an absolute and a relative state of nature. Aquinas identified the relative state of nature with the human condition after the fall of Adam and Eve, which grounds a transition from innocence to sinfulness and the consequent weakening of human nature with its disastrous personal and social implications. Thus, it became necessary to introduce government and laws to regulate community and to restrain evil. Aquinas inserted these views of nature into a two-story model of the universe, the supernatural constructed on top of the natural. The first story is the created world—the world of nature—which has passed from innocence to sin and corruption. The second story is the supernatural—the world of God and the heavenly court—which is the goal and fulfillment of the natural world. The supernatural provides the purpose for which the natural was created. These two worlds are united by two major social realities: law and the institutions of church and state.

Eternal law for Aquinas is the plan or order of creation that has existed in the mind of God from all eternity. It is a divine reality, immutable, and the locus of all truth and values. Consequently, it is the source of all morality. Parallel to this is the natural law, the apprehension and articulation of this eternal law by human reason. Natural law for Aquinas finds its roots in the Stoic absolute law of nature; but given his take on human weakness and the clouding of the intellect due to sin, the natural law of Aquinas is actually closer to the Stoic relative law of nature. In addition, Aquinas defines two other types of law: divine law—that is, the law of the church derived from revelation but also having affinities to natural law and human—and governmental law, which regulates personal and social interaction. This division of law forms the basic matrix within which all persons and institutions act and provides the norms to which they must conform.

In the Thomistic synthesis, the person stands in the center of two intersecting lines, the supernatural and the natural, which are united through eternal law and natural law plus the laws of the church and the state. Consequently, Aquinas was able to construct a universe that was organically united politically and theologically, with a transcendent norm, a norm by which one could evaluate objectively both political and religious behavior. This model of the universe produced an organic unity of civilization in which there was a proper ordering of all things and harmony not only within each separate order but also between the different orders.

Of particular importance in this theory, particularly as interpreted and practiced by the magisterium, is the status of physical acts. This perspective, called physicalism, emphasizes "specifically physical and biological aspects of actions and their consequences."[1] This emphasis, which gives a lesser role to other aspects of human acts such as the social, relational, or psychological, "is based partly on a static and essentialist definition of human nature."[2] More specifically, in physicalism,

> the emphasis is placed on the physical *finis operis* (end of the act), *objectum*, or actual physical properties, motions, and goals of the action under consideration. Within a static natural understanding of human nature, the ethical judgments arrived at are considered to be universally applicable to all situations involving the same physical act. Physicalist criteria are used to determine the *finis operis* (the "end" of the "act-in-itself") or the *objectum*.[3]

What is of critical significance is the moral link between a particular act and the moral order. This link is based on the metaphysical assertion that the biological order reflects the eternal law in the mind of God and constitutes the foundation for an objective moral order.

Two examples will clarify the issue. In a 1944 address to the Italian Medical-Biological Union of St. Luke, Pius XII said:

> In forming man, God regulated each of his functions, assigning them to the various organs. In this way, He distinguished those which are essential to life from those which contribute only the integrity of the body, however precious be the activity, well-being, and beauty of this last. At the same time, God fixed, prescribed, and limited the use of each organ. He cannot therefore allow man now to arrange his life and the function of his organs according to his own taste, in a manner contrary to the intrinsic and immanent function assigned them.[4]

More recently, Pope John Paul II, writing as Cardinal Karol Wojtyla, stated the position this way:

> Thus, the whole order of nature has its origin in God, since it rests directly on the essences (or natures) of existing creatures from which arise all dependencies, relationships and connections between them. In the world of creatures inferior to man, creatures without reason, the order of nature is realized through the workings of nature itself, by way of instinct and (in the animal world) with the help of sensory cognition. In the world of human beings the dictates of the natural order are realized in a different way—they must be understood and rationally accepted. *And this understanding and rational acceptance of the order of nature—is at the same time recognition of the rights of the Creator.* Elementary justice on the part of man towards God is founded on it. Man is just towards God the Creator when he recognizes the order of nature and conforms to it in his actions.[5]

This understanding of morality is present throughout the teachings of the popes and various congregations. It assumes a static order of reality in which certain physical functions and acts have been assigned a role by the Creator and morality consists in knowing these functions and conforming oneself to this order established by God.

Magisterial Statements

To construct the framework in more detail, we now turn to specific magisterial statements on moral law, human reproduction, and genetics.

POPE PIUS XII.   Presented here are themes from Pope Pius XII that are relevant for our review. The first has to do with nature and the law of nature. Pius specified an order created by God that is "directed to the end designed by the Creator. It embraces not only the external acts of man, but also the internal consent of his free will."[6] This order sets a limit on what the individual can do: "Man, in truth, is not the owner of his body nor its absolute lord, but only its user."[7] This natural order governs what can be done to or with the body. In another context, Pius XII expressed the ethic of stewardship this way:

> As far as the patient is concerned, he is not absolute master of himself, of his body, or of his soul. He cannot, therefore, freely dispose of himself as he pleases. Even the motive for which he acts is not by itself either sufficient or determining. The patient is bound by the immanent purposes fixed by nature. He possesses the right to use, limited by natural finality, the faculties and powers of his human nature.[8]

Finally, in speaking of genetics, Pius noted that "where man is concerned, genetics is always dealing with personal beings, with inviolable rights, with individuals who, for their part, are bound by unshakable moral laws, in using their power to raise up a new life. Thus the Creator Himself has established certain barriers in the moral domain, which no human power has authority to remove."[9] Pius XII clearly and forcibly affirmed the presence of an order created by God that, on the one hand, is the basis of the moral order and, on the other hand, sets limits to human activity. This moral order stands independent of human desire or intention. It provides the objective basis for knowing what is good or evil, right or wrong.

On this basis Pius XII condemned the eugenics movement, all forms of artificial insemination, in vitro fertilization, direct sterilization, and the prohibition of marriage to those with genetic illness. In short, Pius XII, on the basis of natural law, prohibited any interference with the physical act of heterosexual married intercourse. The morality of heterosexual married intercourse—as well as any study or use of genetic material—is based on the biological integrity of the physical act of intercourse. Any attempt at interference is considered stepping outside the bounds imposed by nature and constitutes a violation of the stewardship ethic.

Pius XII obviously did not speak directly of genetic engineering or genetic research. Such was not a possibility during his life at midcentury. Yet the seeds of such a response are present, and they strongly point to a conservative moral position.

VATICAN II. The Second Vatican Council (1962–65) also contributed to the contemporary Catholic framework. Vatican II noted that the mastery of humans has extended over almost the whole of nature. In itself this type of activity is part of God's will: "For man, created to God's image, received a mandate to subject to himself the earth and all that it contains, and to govern the world with justice and holiness."[10] The council drew this conclusion: "Hence it is clear that men are not deterred by the Christian message from building up the world, or impelled to neglect the welfare of their fellows. They are, rather, more stringently bound to do these very things."[11]

The council provided ethical guidelines for this building up of the world. For example, the council proposed as the norm of human activity "that in accord with the divine plan and will, it should harmonize with the genuine good of the human race, and allow men as individuals and as members of society to pursue their total vocation and fulfill it."[12] This is further specified by the assertion that by the very fact of being created, "all things are endowed with their own stability, truth, goodness, proper laws, and order. Man must respect these as he isolates them by the appropriate methods of the individual science or arts."[13]

Vatican II was walking a fine line. On the one hand, *Gaudium et Spes* asserts strongly the legitimate autonomy of created reality. "We cannot," the council declared, "but deplore certain habits of mind, sometimes found too among Christians, which do not sufficiently attend to the rightful independence of science."[14] But such autonomy does not mean that created things do not depend on God or that "man can use them without any reference to the Creator."[15] Thus, there is an affirmation of natural law, but here it is conditioned somewhat by the emphasis on the integrity of created reality.

A tension is shown clearly when the *Gaudium et Spes* document discusses human reproduction: "Therefore when there is question of harmonizing conjugal love with the responsible transmission of life, the moral aspect of any procedure does not depend solely on sincere intentions or on an evaluation of motives. It must be determined by objective standards."[16] As such, this statement is close to the teaching of Pius XII. But then the text goes on to say that these standards must be "based on the nature of the human person and his acts."[17] This part of the criterion, though rooted in the tradition, opens the way to a consideration of the person that incorporates more than the biology of his or her acts.

Again no specific statement on genetics is made, but one paragraph will provide an orientation to various topics in the years to follow:

> For God, the Lord of life, has conferred on men, the surpassing ministry of safeguarding life—a ministry which must be fulfilled in a manner which is worthy of man. Therefore from the moment of its conception life must be guarded with the greatest care, while abortion and infanticide are unspeakable crimes. The sexual characteristics of man and the final faculty of reproduction wonderfully exceed the dispositions of lower forms of life. . . . Relying on these principles, sons of the Church may not undertake methods of regulating procreation which are found blameworthy by the teaching authority of the Church in its unfolding of the divine law.[18]

POPE PAUL VI. Pope Paul VI gave further specification to the content of the law of nature and the ethics of stewardship in the encyclical *Humanae Vitae* in 1968. Although its focus is the ethics of heterosexual married reproduction, the argument on which the encyclical is based is critical in understanding many of the themes of this chapter.

Critical to the argument of the encyclical is the notion of responsible parenthood. This concept has three levels of meaning. First, the biological discussion "means the knowledge and respect of their functions; human intellect discovers in the poser of giving life biological laws which are part of the human person."[19] Second is the instinctual, which means "that necessary dominion which reason and will must exercise over them [instinct and passion]."[20] Third, there is the socioeconomic dimension, which either accepts the children born to the family or, for a time and in accord with the moral law, avoids a new birth.

Critical to the argument is the "inseparable connection, willed by God and unable to be broken by man on his own initiative, between the two meanings of the conjugal act: the unitive meaning and the procreative meaning."[21] Such a structure "capacitates them for the generation of new lives, according to laws inscribed in the very being of man and of woman."[22]

Of critical importance in developing the ethic surrounding human reproduction is the comment

> to make use of the gift of conjugal love while respecting the laws of the generative process means to acknowledge oneself not to be the arbiter of the sources of human life, but rather the minister of the design established by the Creator. In fact, just as man does not have unlimited dominion over his body in general, so also, with particular reason, he has no such dominion over his generative faculties as such, because of their instinctive ordination toward raising up life, of which God is the principle.[23]

This affirmation of an inherent limit to human action based on the physical structure of the person is further specified later in the encyclical:

> Consequently, if the mission of generating life is not to be exposed to the arbitrary will of men, one must necessarily recognize insurmountable limits to the possibility of man's domination over his own body and its functions; limits which no man, whether a private individual or one invested with authority may licitly surpass. And such limits cannot be determined otherwise than by the respect due to the integrity of the human organism and its functions, according to the principles recalled earlier, and also according to the correct understanding of the "principle of totality" illustrated by our predecessor Pope Pius XII.[24]

Paul VI clearly invoked a particular understanding of the nature of the human person and a standard of morality based on that as a way of identifying the limits to human activity and dominion. Biological structures morally qualify human activity since disregarding them violates the order willed by God in the creation of human beings.

POPE JOHN PAUL II. We will now survey samples of the teaching of John Paul II. Since his comments are numerous and more detailed analysis is available elsewhere, only representatives of his teachings will be presented.

In his book *Love and Responsibility*, written while he was still Cardinal Karol

Wojtyla, he reveals a vision that sets the framework for his further thought: the heart of the moral order has its origin in justice to God and is based on the fact of creation. The order of nature has its origin in God "since it rests directly on the essences (or natures) of existing creatures, from which arise all dependencies, relationships and connections between them."[25] The order of nature grounds morality.

> In the world of human beings the dictates of the natural order are realized in a different way—they must be understood and rationally accepted. *And this understanding and rational acceptance of the order of nature—is at the same time recognition of the rights of the Creator.* Elementary justice on the part of man towards God is founded on it. Man is just towards God the Creator when he recognizes the order of nature and conforms to it in his actions.[26]

Or, as John Paul II again states it: "But before and above all else man's conscience, his immediate guide in all his doings, must be in harmony with the law of nature. When it is, man is just towards the Creator."[27] Additionally, this view makes the human an active participant in the order of nature for, in a classic Thomistic statement of the issue, "Man, by understanding the order of nature and conforming to it in his actions, participates in the thought of God, becomes *participes Creatoris*, has a share in the law which God bestowed on the world when he created it at the beginning of time."[28]

The pope distinguishes between the biological order and the order of nature. The order of nature is the "divine order inasmuch as it is realized under the continuous influence of God the Creator."[29] The biological order means the "same as the order of nature but only in so far as this is accessible to the methods of empirical and descriptive natural science."[30] It is not a specific order of existence with its own relation to God but a physical description of the natural order created and structured by God.

Within the natural order, the human being has a structure. On the one hand, humans are "only required not to destroy or squander those natural resources, but to use them with restraint, so as not to impede the development of man himself, and so as to ensure the coexistence of human societies in justice and harmony."[31] On the other hand, "the sexual urge is connected in a special way with the natural order of existence, which is the divine order."[32] And that determines the proper purpose of the sexual urge. Thus, "it is not the love of man and woman that determine the proper purpose of the sexual urge. The proper end of the urge, the end per se, is something supra-personal, the existence of the species Homo, the constant prolongation of its existence."[33] Such a respect of the order of nature sets out the following perspective then:

Marriage, objectively considered, must provide first of all the means of continuing existence, secondly a conjugal life for man and woman, and thirdly a legitimate orientation for desire. The ends of marriage, in the order mentioned, are incompatible with any subjectivist interpretation of the sexual urge, and therefore demand from man, as a person, objectivity in his thinking on sexual matters, and above all in his behavior. This objectivity is the foundation of conjugal morality.[34]

Morality in general and in marriage in particular consists in active conformity to the order of nature through which humans participate in the plan of God for creator and become participants in creation. Two conclusions follow from this. First, with respect to marriage: "When couples, by means of recourse to contraception, separate these two meanings [the unitive and procreative] that God the Creator has inscribed in the being of man and woman in the dynamism of their sexual communion, they act as 'arbiters' of the divine plan and they 'manipulate' and degrade human sexuality—and with it themselves and their married partner—by altering its value of 'total' self-giving."[35] Second, with respect to public policy: "Thus the Church condemns as a grave offense against human dignity and justice all those activities of human governments or other public authorities, which attempt to limit in any way the freedom of couples in deciding about children. Consequently, any violence applied by such authorities in favor of contraception or, still worse, of sterilization and procured abortion, must be altogether condemned and forcefully rejected."[36]

Finally, Pope John Paul II condemns "in the most explicit and formal way experimental manipulations of the human embryo, since the human being, from conception to death, cannot be exploited for any purpose whatsoever."[37] Thus, the criterion of the exploitation of the human imposes a clear limit on experimentation.

THE CONGREGATION FOR THE DOCTRINE OF THE FAITH. The Congregation for the Doctrine of the Faith has also issued a major statement regarding human reproduction, *Donum Vitae*. This document also relies on the natural law tradition as the foundation of its moral analysis. The natural moral law is defined as "the rational order whereby man is called by the Creator to direct and regulate his life and action and in particular to make use of his own body."[38] Two norms particularly identified for ethical analysis are the life of the human being and the special nature of the transmission of life in marriage.

The statements about human life have several dimensions. First is the affirmation that "from the moment of conception, the life of every human being is to be respected in an absolute way because man is the only creature on earth that God has 'wished for himself' and the spiritual soul of each man is 'immediately created' by God."[39] Second, since God is the Sovereign of life from beginning to its end, "no one can in any circumstance claim for himself the right to destroy directly an innocent human being."[40] Third, an intervention "on the human body affects not only the tissues, the organs and their functions, but also involves the person himself on different levels. It involves, therefore, perhaps in an implicit but nonetheless real way, a moral significance and responsibility."[41]

With respect to the transmission of human life, *Donum Vitae* affirms

> the "inseparable connection, willed by God and unable to be broken by man on his own initiative, between the two meanings of the conjugal act: the unitive meaning and the procreative meaning. Indeed, by its intimate structure the conjugal act, while most closely uniting husband and wife capacitates them for the generation of new lives according to laws inscribed in the very being of man and of woman." This principle, which is based upon the nature of marriage and the intimate connection of the goods of marriage, has well-known consequences on the level of responsible fatherhood and motherhood. "By safeguarding both these essential aspects, the unitive and the procreative, the conjugal act preserves in its fullness the sense of true mutual love and its ordination toward man's exalted vocation to parenthood."[42]

Because of the moral significance of the links between the conjugal act and the goods of marriage, the unity of the person and the dignity of his or her origin, "*the procreation of a human person [must] be brought about as the fruit of the conjugal act specific to the love between spouses.*"[43] Therefore, "no one may subject the coming of a child into the world to conditions of technical efficiency which are to be evaluated according to standards of control and dominion."[44]

While the implications of this for all sorts of artificial reproduction are obvious, these two principles also speak to the issues of genetic testing and fetal research. Prenatal diagnosis is morally licit if it "respects the life and integrity of the embryo and the human fetus and is directed toward its safeguarding or healing as an individual."[45] Any link of prenatal diagnosis to direct abortion is prohibited. Also prohibited are any professional, social, or scientific linkages between prenatal diagnosis and abortion. Thus, prenatal diagnosis is morally permissible "with the consent of the parents after they have been adequately informed, if the methods employed safeguard the life and integrity of the embryo and the mother, without subjecting them to disproportionate risks."[46]

Therapeutic procedures on the embryo are morally permissible if they "*respect the life and integrity of the embryo and do not involve disproportionate risks for it, but are directed toward its healing, the improvement of its condition of health or its individual survival.*"[47] Research or experimentation can be carried out on embryos and fetuses if "*there is a moral certainty of not causing harm to the life or integrity of the unborn child and the mother, and on condition that the parents have given their free and informed consent to the procedure.*"[48] Additionally, living embryos, viable or not, "*must be respected just like any other human person; experimentation on embryos which is not directly therapeutic is illicit.*"[49]

Embryos may not be produced in vitro to be used for experimentation nor may embryos obtained in this way for research be destroyed. The reason is that "by acting in this way the researcher usurps the place of God; and, even though he may be unaware of this, he sets himself up as the master of the destiny of others inasmuch as he arbitrarily chooses whom he will allow to live and whom he will send to death and kills defenseless human beings."[50]

BISHOPS OF GREAT BRITAIN.   In 1990, the Catholic Bishops' Joint Committee on Bioethical Issues of Great Britain issued a statement entitled "Antenatal Tests: What You Should Know."[51] This document provides an ethical analysis of several prenatal tests that women may be routinely offered. Two lines of reasoning are put forward.

First is an affirmation of the equal right to life of each person, which includes the fetus since "there is no stage of growth at which the developing being is not the same being as the adult individual. To cut short this life is to cut short the life of a person."[52] Since each life is a value in itself, it is an indignity to speak of disabled persons as being burdens on others. Such value is also the basis for rejecting arguments for abortion based on quality of life, particularly since "a life which from a healthy person's point of view seems frustrating, miserable and unfulfilled, may be viewed very differently by the person whose life it is."[53] Such quality of life arguments deny the equal dignity of each person. Therefore, the bishops of Great Britain argue that it is not right to kill a disabled unborn child: "The child's plight is, instead, a call for both justice and charity. True charity seeks to relieve the suffering of others, but not by putting an end to the lives of afflicted individuals."[54]

Second, the bishops affirm that "blood tests in early pregnancy and ultrasound scans to establish that the mother's dates are right and the child is developing well, and to find out whether she is pregnant with one or more babies"[55] are quite legitimate. The blood test for serum alphafetoprotein, which detects spina bifida, should not be accepted by the mother, for its only purpose is to detect spina bifida "to give the mother the choice to terminate the pregnancy if the baby is affected. For mothers who do not contemplate abortion it can serve no purpose."[56] In speaking about amniocentesis, chorian biopsy, and fetoscopy, the bishops argued that "it is wrong for a pregnant mother to have any of these tests in order to decide whether or not to keep her baby. But a test is morally justified if its purpose is to find out how best to look after the baby during pregnancy."[57] Prenatal testing has a place in the management of a pregnancy, but that use is limited to promoting the baby's health.

THE CATHOLIC HEALTH ASSOCIATION. Finally, we review the document *Human Genetics: Ethical Issues in Genetic Testing, Counseling, and Therapy* published by the Catholic Health Association of the United States (CHA). Although it is not official magisterial teaching, I include this document because of CHA's quasi-official standing.

After providing an overview of genetic diseases and noting that "many equate the providing of information with moral approval or recommendations of particular options,"[58] the document then moves into an analysis of various ethical issues.

First is an examination of issues in genetic testing. The ethical issues are the standard ones: eligibility criteria, truth-telling, whether or not to disclose inadvertent or unanticipated findings, interpreting ambiguous test results, and confidentiality.[59] With respect to somatic cell therapies, CHA argues that they raise no ethical problems other than those already associated with the introducing of innovative therapies or research protocols. Germ-line therapy does introduce several ethical problems, including eugenic concerns, particularly related to the shaping of future generations; the creation of embryos explicitly for research; and the capacity for preimplantation screening, which appears to make germ-line therapy redundant.[60]

With respect to the genetic counselor, the CHA recognizes that the traditional client-centered or nondirective approach of the profession may raise particular problems for the Catholic counselor or a Catholic institution:

- Criteria for evaluating the moral acceptability of a therapeutic intervention are identified.
- Human dignity must be respected and thus human identity safeguarded.
- The intervention must not infringe on the origin of human life; that is, procreation linked to both biological, as well as spiritual, union of the parents united in matrimony.
- The intervention will avoid manipulations that tend to modify genetic inheritance and to create new groups at risk of marginalization in society.
- Racist and materialist attitudes do not inspire the intervention.[61]

Second, ethical guidelines for counseling are proposed: "While autonomy is to be respected, individual persons must be aware that they have an obligation to use reasonable means to prevent harm to others."[62] Also critical here is the affirmation that individuals or couples should not be unduly coerced by social pressures to use or not to use genetic services: "The promotion of personal well-being is a primary ethical goal of all medical practice."[63] A particular nuance here is to "help the client interpret well-being in the context of family relation-ships, and thus treat each person so as to strengthen family well-being."[64] Dis-closure and confidentiality are vital. Information should be "provided promptly, in a form that is clear, understandable, and relevant to the client."[65] Addition-ally, information "is exchanged in genetic counseling with the understanding that it will be kept private."[66] Nonetheless at times disclosure may be justified. Five criteria are identified:

- There is strong probability both that harm will occur if the informa-tion is withheld and that the disclosed information will actually be used to avert harm.
- Reasonable efforts to elicit voluntary consent to disclosure have failed.
- Identifiable persons would suffer serious harm without disclosure.
- The information will actually be used to avert harm.
- Only the information necessary to prevent harm is disclosed.[67]

Finally, the CHA document argues that Catholic institutions should provide genetic counseling services. A fourfold argument is developed:

- Genetic services are available elsewhere. If Catholic institutions do not offer them, clients will seek them in other centers.
- These services will continue to be provided whether Catholic institu-tions provide them or not.
- Other institutions or centers have less or no problem with certain pro-cedures, such as abortion.
- A Catholic center, being opposed to such procedures as abortion, will provide counseling and support for morally acceptable alternatives.[68]

Thus, the CHA affirms the moral context of genetic counseling, specific moral problems associated with certain procedures, but argues that the Catholic coun-seling center can provide appropriate moral support for those seeking genetic information and, indeed, Catholic institutions ought to offer such services.

Theological Commentary Resources

This section reviews the writings of several theologians on the topic of genetic engineering and genetic screening. The purpose of this section is to move from frame to focus—that is, to identify ideas and themes in the theological litera-ture that will carry the analysis of the ethical issues forward.

Karl Rahner, S.J.

Perhaps the reigning voice in twentieth-century Roman Catholic theology, Karl Rahner has had much to say about human nature and human self-creation that is relevant to our theme. He also spoke directly to the topic of genetic engineering.

First let us examine his perspectives on human nature. Christian anthropology defines the human being as a self-creating being, a "free being before God, a person subject to himself and capable of freely determining his own final condition. This self-determination is so complete that he can ultimately and absolutely become what he wants to be."[69] As created, the human is essentially unfinished, and one determines oneself through free action. What is new is how the power of self-creation "has now grasped the physical, psychological, and sociological dimensions of his existence."[70] Now we have become free beings under our "own responsible control."[71] And this self-creation will "develop the concrete form of human openness which leads to the absolute future that comes from God."[72] However, this self-creation can have irreversible consequences. Christian anthropology puts us on guard against the dangers involved: "The fall initiated a process that allowed for no return."[73]

Such a self-creation proceeds from respect for nature and yet helps form human nature. Critical for Rahner is the fact that the nature that must be respected "is a nature still being actively formed by man himself through self-creation."[74] Human nature is not something fixed or given. Our understanding of human nature will be experienced in no "other way than in a particular historical form, where we are unable to distinguish clearly and unequivocally between the nature 'in itself' and its concrete variable form[?]."[75]

A critical counterpoint to the theme of self-creation is Rahner's insistence that there is another radical *existentiale* that sets a limit to planning: the human "is no less a being whose essence has been predetermined."[76] That is, the human has not called himself or herself into existence. We have not chosen our world but have "been projected into a particular world and . . . this world is presented to [us] for [our] free acceptance."[77] The very important conclusion—and basis of the limit of self-creation—is "the world can never be 'worked over' to such an extent that man is eventually dealing only with material *he* has chosen and created."[78] For Rahner, then, a necessarily given factor of human nature is "accepting [that] this necessarily alien determination of one's own being *is* and *remains*, therefore, a fundamental task of man in his free moral existence."[79] In concrete terms this means that "genetic manipulation is the embodiment of the fear of oneself, the fear of accepting one's self as the unknown quantity it is."[80] The driving force behind genetic manipulation, from this perspective of the limit on our nature, is "despair because he cannot *dispose* of existence."[81]

Therefore with the Rahnerian perspective, there is a critical tension. On the one hand, the human is called to self-creation in radical freedom through which act the absolute future is encountered. On the other hand, there is a clear limit to self-creation: the radical givenness of the human condition that prohibits a radical disposal of ourselves. This leads to the rather somber conclusion that "if the new humanity of the future is to survive, it must cultivate a sober and critical resistance to the fascination of novel possibilities."[82]

Rahner discusses two genetic manipulations. First is experimentation with human genetic material. After distinguishing between obtaining the material and the actual experiment itself—and arguing that obtaining such material is not necessarily problematic—Rahner turns his attention to the experiment itself. In responding, he makes a critical observation. If, as the tradition presupposes, a human being comes into existence at conception, we have an individual person with his or her own rights: "*If* this is the case, such a person is no more an inconsequential passive object for experiments than the prisoners of Nazi concentration camps."[83] However, this presupposition "is exposed to positive doubt."[84] Although strongly rejecting the conclusion that such a doubt does not reduce the material to a thing, Rahner nonetheless argues, "It would be conceivable that, given a serious positive doubt about the human quality of the experimental material, the reasons in favor of experimenting might carry more weight, considered rationally, than the uncertain rights of a human being whose very existence is in doubt."[85]

Second, Rahner turns his attention to heterologous artificial insemination. In this context he also speaks of in vitro fertilization, and his comments are a critique of both. Rahner argues against these interventions because this manipulation "fundamentally separates the marital union from the procreation of a new person as this permanent embodiment of the unity of married love; and it transfers procreation, isolated and torn from its human matrix, to an area outside man's sphere of intimacy."[86] Such a manipulation reflects a desire to plan the human totally and, as such, represents a transgression of the *existentiale* that rejects our total disposability. Rahner sees a most compelling social reason for the prohibition of such manipulation: "To pursue the practical possibility of genetically manipulating man is to threaten and encroach upon this free area. For it offers incalculable opportunities of man's manipulation—reaching to the very roots of his existence—*by organized society*, i.e., the state."[87]

Rahner argues that the danger of misusing a new possibility may be taken into account if the new possibility is justified, yet he does not seem too sure about the possibility of such justification: "What is the point of genetic manipulation if not to extend the state's area of control and thus to diminish, instead of to increase, man's sphere of freedom?"[88] He argues that since the new possibilities can be discovered faster than their effects on humanity ascertained, "it is so vital for humanity to develop a resistance to the fascination of novel possibilities."[89] Such a critical analysis comes from both the reality of the Fall and the limit on our capacity for self-creation: the givenness of our own nature.

Bernard Häring, C.S.S.R.

Ethicist Bernard Häring presents his comments on genetic engineering in the context of the wider theme of the ethics of manipulation. In his view the ethics of manipulation—the use of someone or thing—is set by the worldview of its users or by a technological view or view of wisdom that includes "the ability to reciprocate genuine love."[90] For Häring the problem is not manipulation per se: "The evil is in the transgression of the limits posed by human freedom and dignity."[91] Therefore, the formation of a pure technocrat—*homo faber*—"is formed to the detriment of *homo sapiens*, the discerning and loving person."[92] Another dimension of his critique is that "behind many individual acts of manipulation that degrade the freedom and dignity of persons, stands the ideology that equates technical progress with human progress."[93] Although the individual that Häring calls technological man is legitimately a manipulator and indeed has a right and duty to manipulate for the benefit of humanity, this individual "is a slave to ideology if he measures everything by criteria appropriate only to the field of technology."[94]

A clear vision of human nature informs Häring's vision:

> The fundamental condition for being truly free while acting as manipulator of the world around is our sabbath, our repose before God. Only if man transcends himself and recognizes the gratuity of all creation and of his own call to be a co-creator, can he submit the earth to his own dignity.[95]

This foundation gives us the ability to keep a sense of admiration and adoration, without which, Häring says, "our manipulation of the world becomes depletion and alienation."[96] The vocation of becoming a co-creator and co-revealer through uniting with the human community is to enhance the "freedom and co-responsibility of all mankind."[97] In so transforming the world, the person does not act merely as a consumer or manipulator. Rather, the person is "an artist, and he grows in awareness and dignity while transforming the work, provided he sees his highest creativity in mutual respect and reverence in all his human relations."[98]

In Häring's vision, freedom plays a large role. This inner personal freedom is the "capacity to long for ever-growing knowledge of what is good and truthful, the capacity to love what is good and put it into practice."[99] Critical elements of this are our own "self-interpretation, self-awareness, and his or her active insertion into the history of liberation."[100] This becomes a critical touchstone in assessing the ethics of a particular manipulation.

Related to freedom is the concept of stewardship. Häring interprets stewardship in the light of the human's noblest vocation: the capacity to "freely interfere with and manipulate the function of his *bios* and psyche in so far as this does not degrade him or diminish him or his fellowmen's dignity and freedom."[101] Both nature and the person's nature call "for his free stewardship, his creative cooperation with the divine artist."[102] For Häring we are, under God, our own providence and have a right and duty to plan our future and pilot evolution. But in doing this, our first concern should be "to explore the true dimensions of freedom and never to take the risk of diminishing or losing our own freedom or jeopardizing the freedom of others."[103] In planning a particular manipulation, we need to think of both the means and the end: "If a concrete form of manipulation violates the basic values of respect for human freedom and dignity, or other equally high values, then the hoped-for consequences cannot justify the means."[104] Such a need to evaluate the means and the consequences comes from the fact that as humans, we are an unfinished and indeed an uneven world. "We bear in our genes, in the millions of data stored in our brains and in our environment, the burden of the sins of many generations and many people."[105]

In turning to genetic engineering, Häring situates the ethical context this way: "For Christians, the theological themes of creation and procreation bring genetics into the field of ethics. The divine mandate to submit the earth and to fill it includes man's mission to transform life according to his finest vision of humankind's future."[106] The main problem that he sees is "a one-sided technological thinking which either cannot or will not face ethical values and standards."[107] The critical ethical question is what kind of person do we desire: "the highly qualified technical and perhaps emotionally and morally undeveloped man, or the human person with, perhaps, a less developed IQ when measured for technical capacities, but more human, more able to grow in love and to discern with wisdom what is truly love, inner freedom, generosity, and so on"?[108]

With respect to particular interventions, Häring is fairly specific. Premarital genetic screening is highly desirable, and a premarital exchange of genetic information, a basic moral duty. Genetic surgery—if effective and having no disproportionate risks—is the most moral and promising genetic intervention. Häring sees it difficult to justify amniocentesis unless it is done with a view to therapy.

With respect to genetic engineering—the manipulation of the genome per se—Häring identifies three questions: "(1) Is mankind allowed to try, by direct gene manipulation, to improve the human species beyond the indication of therapy? (2) If so, can we offer criteria? (3) Can we have any trust that the technical man of today will approach such a daring enterprise in the right spirit?"[109]

For Häring, the careful planning of constructive changes to improve the species "cannot be rejected *a priori* as being against man's nature or vocation."[110]

Yet he argues that there are substantial reasons to "fear that genetic engineering could fall under the heartless rules of the market."[111] In summary, Häring argues that we have a limited right of self-modification, though "it is no easy matter to determine accurately the legitimate limits of such a new venture."[112] Initial criteria for discerning the limits are human freedom, the capacity for intercommunication, and an affirmation of human wisdom, as opposed to technical intelligence only.

Häring concludes by arguing that our new capacities mandate that we make good use of our knowledge:

> To that end we all have to learn to discern better what genuine therapy and human progress mean in a perspective of human dignity and freedom. Those who belittle, ignore, or plainly deny these basic values of dignity and freedom have, willingly or unwillingly, called for the animal arising from the abyss. It is urgent, therefore, that we confront him now, before he is fully aroused. Once aroused, as history has shown, we will demand the holocaust.
>
> We need, above all, to hold fast to our sense of mystery, our capacity for admiration, for celebration, for contemplation, and for a wholehearted common search for ultimate values.[113]

Richard A. McCormick, S.J.

This section presents a small angle of vision on Richard A. McCormick's perspective on genetic engineering. The task is made a little more difficult because in a wondrous way McCormick's ideas keep developing. Any sample of his writing is necessarily limited and tentative.

McCormick has provided a general overview of critical ethical themes that can serve as a context in which to present his own analysis. He has six themes that ought to inform our bioethical reasoning.

First is life as a basic but not absolute good.[114] The issue here is that life itself can cede other values: the glory of God, the salvation of souls, or service to one's neighbor. Second, a vision of the value of life must inform Christian reflections about nascent life and, therefore, a "simple pro-choice *moral* position is in conflict with the biblical story."[115] Third, in the Christian perspective, "the meaning, substance, and consummation of life are found in *human relationships*."[116] Life is to be valued as a condition for other values, and since these values cluster around human relations, life is a value to be preserved "only insofar as it contains some potentiality for human relationships."[117] Fourth, our well-being is pursued only interdependently. Sociality is an essential part of both our being and our becoming. Fifth, the spheres of life giving and love making are not to be radically put asunder. McCormick's emphasis here is on the spheres of the relationship as a whole, not the isolated acts of a couple. Sixth is "heterosexual, permanent marriage as normative."[118] By this McCormick means that "monogamous marriage provides us with our best chance to humanize our sexuality and bridge the separateness and isolation of our individual selves."[119]

This gives the general context for presenting specific analysis of McCormick's discussion of genetic engineering. These six themes do not exhaust McCormick's thinking on these topics, but they are themes to which he continually refers as touchstones for his analysis.

With respect to amniocentesis, McCormick follows the framework suggested above by the CHA to justify offering amniocentesis at Catholic institutions: in the context of Catholic teaching, amniocentesis in Catholic institutions will save fetal lives that would be lost. But McCormick notes that, with the procedure, the institutions must also implement a support system for the problem pregnancy. If they do not, the reason for amniocentesis is gone, and they "are simply part of a system making abortion more likely."[120] Such a policy must be implemented prudently, though, because it must balance between respect for the fetus and respect for the conscience of the parents, between compassion and coercion.

Regarding the moral status of the embryo, McCormick argues that since "there are significant phenomena in the preimplantation period that suggest a different evaluation of human life at this stage . . . I do not believe that nascent life at this stage makes the same demands for protection that it does later."[121] In part this evaluation depends on a differentiation between genetic individuation and developmental individuation. Even though one's individuality (except in the case of twinning) may be fixed with the genetic code at fertilization, becoming a person is a developmental process. One cannot with certainty draw the conclusion that one is a person at the preimplantation stage, but one could draw the conclusion that one should be treated as a person.

With regard to genetic engineering, McCormick comments on several specific interventions. He, as with others, states that somatic cell therapy—altering the function of a defective gene—should be understood as "nothing more than an extension of medical practice in an attempt to aid victims of currently intractable diseases."[122] The relevant ethical criteria for this procedure are effectiveness and safety. With respect to enhancement genetic engineering—the attempt to produce a desired characteristic—McCormick identifies two ethical criteria. First, the possibility of the gene's affecting a nontargeted function in a healthy human. Second, a shift in valuing humans not for the whole that they are but "for the *part* that we select."[123] Finally regarding eugenic genetic engineering—systematic preferential breeding of superior individuals—

McCormick observes, "Ethically, the matter is quite straightforward, and it is all bad."[124] What characteristics, he asks, are to be preferred, and who is to decide all this?

Two criteria are proposed for evaluating genetic engineering. The first, derived from Vatican II's standard of the nature of the person and the person's acts, asks, "Will this or that intervention (or omission, exception, policy, law) promote or undermine human persons 'integrally and adequately considered'?"[125] Second are four specific values: the sacredness of human life; the interconnection of life systems; individuality and diversity; and social responsibility and the priorities of research.[126] These values set a context for genetic engineering that forces us to look beyond a narrow calculation of individual risks and benefits and to take account, as best we can, of the context in which genes will be engineered.

On a policy level, McCormick argues against experimentation on embryos, with exceptions "allowed only after scrutiny and approval by an appropriate body."[127] The fact of doubt about the status of personhood of the early embryo leaves open the possibility of the use of such an entity or its tissue in some research. Thus, one can conclude that McCormick leaves us with a carefully guarded openness to some genetic interventions.

Charles Curran

Like McCormick, Charles Curran has written frequently and widely, and providing a full overview of his thought is beyond the scope of this chapter. Only representative areas of his thought will be presented. Curran's moral method "consists of a perspective based on the fivefold Christian mysteries of creation, sin, incarnation, redemption, and resurrection destiny."[128] This stance provides a positive method and perspective.

> Creation indicates the goodness of the human and human reason; but sin touches all reality, without, however, destroying the basic goodness of creation. Incarnation integrates all reality into the plan of God's kingdom. Redemption as already present affects all reality, while resurrection destiny as future exists in continuity with the redeemed present but also in discontinuity because the fullness of the kingdom remains God's gracious gift at the end of time.[129]

Curran's model of the Christian life is relationality-responsibility, which views "the moral life primarily in terms of the person's multiple relationships with God, neighbor, world, and self and the subject's actions in this context."[130] It reflects the Christian emphasis on covenant and love. It highlights the need to respond to the contemporary situation and make the dominion more present. And it opens up different approaches to particular questions. Curran develops a Christian anthropology that calls attention to the person in a twofold way: "First, individual actions come from the person and are expressive of the person. . . . Second, the person, through one's actions, develops and constitutes oneself as a moral subject."[131] Our basic relationship to ourselves is governed by "the basic attitude of stewardship, using our gifts, talents, and selves in the living out of the Christian life."[132]

Curran uses a theory of compromise to resolve ethical conflicts. Here Curran refers to "cases in which the presence of sin might justify an action which could not be justified if sin and its effects were not present."[133] Compromise tries to "describe the reality in order to recognize the tension between justifying such actions because of the presence of sin and the Christian obligation to overcome sin and its effects."[134]

With this as background, we now turn to specific issues in genetic engineering. Curran notes that a different anthropology will be required by a contemporary theology, one that recognizes a more open and dynamic understanding of human capacities. "The genius of modern man and woman is the ability toward self-creation and self-direction."[135] But this greater sense of openness must be balanced with a historical approach that prevents an uncritical acceptance of every new intervention as necessarily good. Also in the area of genetics, one must recognize "the existence of other responsibilities which limit one's own options and freedom."[136] Thus, the new anthropology must also avoid an excessive individualism: "My contention is that the complexity and interrelatedness of human existence, plus the tremendous power that science may put into human hands, are going to call for a more communitarian and social approach to the moral problems facing our society."[137]

Curran identifies three dangers in the new genetics against which we must be on guard. First is a "naively optimistic outlook on human growth and progress. . . . Biology or genetics will never completely overcome inherent human limitations and sinfulness."[138] Second is the danger of identifying the scientific with the human, thus forgetting that "the human includes much more than just science and technology."[139] A third danger comes from the success orientation of science and technology, which measures outcomes in terms of results and effects. Such an approach forgets that "the ultimate reason for the lovability of a person does not depend on one's qualities or deeds or successes or failures."[140]

In speaking of the preimplantation embryo, Curran speaks not of a person but of "truly individual human life,"[141] which he argues is present "two to three weeks after conception."[142] This criterion puts a limit on uses of the embryo after that time: "Experimentation after that time and attempts to culture embryos in vitro beyond this stage of development raise insurmountable ethical problems."[143] Curran is struggling with two competing values here: legitimate goals of science in pursuing knowledge and truth plus the value of the embryo. He is fearful that once the research and development of techniques now being perfected on preimplantation embryos are completed, "there will be demands and requests to do research on the embryo after the time of implantation."[144] Curran is concerned about the use of such technologies to attempt to improve the human gene pool. He is "totally opposed to any type of positive eugenics."[145] He wishes to prevent this and proposes two limitations: no use of the embryo after implantation and the limitations of in vitro fertilization to fertility treatments.

Benedict Ashley, O.P., and Kevin O'Rourke, O.P.

In their book *Health Care Ethics*, Benedict Ashley and Kevin O'Rourke present a comprehensive overview of several issues in genetic engineering. They begin with a discussion of ethical issues. They note that the classic way of beginning discussions of genetic engineering is with human dominion over nature. While recognizing the role this dimension has had, Ashley and O'Rourke identify two problems with it. First, it is "too much influenced by the Greek image of God as a jealous monarch who becomes angry when Prometheus infringed on his prerogatives."[146] Alternatively, others see attempts to improve the human being "as an insult to the work of the Creator whose masterpiece is man, or at least as a fatal temptation to pride."[147]

The perspective they propose focuses on three dimensions. First is the fact of God's being a generous Creator who, in creating humans, "called them by the gift of intelligence to share in his creative power."[148] Second is the affirmation of a universe created through an evolutionary process that is not yet complete. Thus, God has made us "co-workers and encourages us to exercise real originality."[149] Third is the fact that our creativity depends on our brains. "Any alteration that would injure the brain and hence a person's very creativity would indeed be a disastrous mutilation, especially if this were to be transmitted genetically."[150] Two criteria, however, allow interventions on other organ systems: "(1) if they gave support to human intelligence by helping the life of the brain and (2) if they did not suppress any of the fundamental human functions that integrate the human personality."[151] Then Ashley and O'Rourke present two general conclusions to this section. Genetic engineering and less radical transformations of the normal human body are permissible "if they improve rather than mutilate the basic human functions, especially as they relate to supporting human intelligence and creativity."[152] Transformations that endanger human intelligence and harm human integrity are prohibited. Second, experiments of such a radical kind must be undertaken with caution and "only on the basis of existing knowledge, not with high risks to the subjects or to the gene pool."[153]

With respect to genetic engineering, or genetic reconstruction as Ashley and O'Rourke call it, of concern is avoiding the reality that all children are loved because they conform to their parents' preferences.[154] While using this as a critique of sex selection technologies, this criterion could also be used as a criterion for a broader critique of particular genetic interventions. Agreeing with the position of many others, Ashley and O'Rourke argue that if the purpose of genetic interventions is individual therapy, the "only ethical issue is the proportion of probable benefit to risk."[155]

With respect to eugenic interventions, particularly producing a human according to a specific profile, for example, height, complexion, or mental abilities, Ashley and O'Rourke argue that "we would not rule it out ethically merely on the grounds that it would be usurpation of God's creative power."[156] Ethical dilemmas arise, however, concerning "whether society has either the knowledge or the virtue to take the responsibility for creating the superior members of the race."[157] They recognize that attempts to define superior are "so ambiguous as to be arbitrary"[158] and propose that rather than define superior by the traits of an age, it should focus on "a being with capabilities of meeting the challenges of new and unpredicted situations."[159] Such challenges will be met by genetic variability, not genetic hybridization.

Ashley and O'Rourke give several broad conclusions about genetic interventions. First, while they recognize the importance of research involving the interaction between genotype and phenotype, priority should be given to research on the phenotype. This could be done primarily by modifications to the environment without direct genetic interventions. Second, currently proposed methods of genetic reconstruction involve in vitro fertilization and other procedures that are "ethically objectionable because they separate reproduction from its parental context and involve the production of human beings. . . . This contravenes the basic principles of ethical experimentation with human subjects."[160] Third, improving the race by selection, cloning, or genetic reconstruction is ethically unacceptable because these methods "restrict the genetic variability important to human survival, and they would separate reproduction from its parental context."[161] Finally, if such mentioned problems can be overcome, "it will be ethically desirable to develop and use genetic methods for therapy of genetic defects in existing embryos, keeping in view the risk-benefit proportion."[162]

Ashley and O'Rourke see the improvement of humans as an exercise of "good stewardship of the share in divine creativity with which God has endowed mankind."[163] But such stewardship should be exercised in such a way that "lest by tampering with their brains or the rest of their personalities they should undermine the freedom and intelligence upon which this creativity depends."[164]

William E. May

Critical for William E. May is the view of human nature one espouses. He is careful to reject inadequate views based on a one-sided scientism that do not take account of the moral nature and freedom of the human being. He is strong in his rejection of the technological imperative that attempts to derive an obligation from a technical capacity. But, as with others, May also recognizes the legitimacy of therapies to prevent or reduce genetic anomalies, though he clearly states that such decisions should not be made on consequentialist or utilitarian grounds.[165]

Experimentation on embryos and cloning, among other technologies, lead to the "voluntary self-degradation and dehumanization of the human beings involved and of the societies of which they are members."[166] Such technologies carry the danger of transforming the home into a laboratory, and reproduction into manufacturing. The results of such technological intervention or experimentation would be products, not persons. However, despite this strong statement, May sees merit in some experiments and research, but argues that the risks clearly need to be taken into account and that the research should be delayed until later in development so that "the genetic malady can be identified more exactly and the possibility of developing specific viruses carrying quite precise types of information in the form of DNA exists."[167]

May argues that premarital genetic counseling ought to be mandatory, for the process [of genetic counseling] "need not carry as its consequence the dehumanization and depersonalization of the transmission of life from one generation to another."[168] As long as a proper vision of the human is maintained and the couples at risk of transmitting genetic diseases "extend their procreative love in other directions,"[169] such interventions can help specific individuals and couples and can help critique eugenic visions of genetics.

This concludes the review of several Roman Catholic theologians with respect to genetic engineering. One can easily see that there is a division between the magisterium and some of the theologians reviewed. At present there seems to be no resolution or grounds for resolution of this division, for they are separated by different methods and anthropologies. The next section does not attempt a resolution of these issues but identifies several themes germane to an ethical and theological assessment of the HGP.

Issues

Many insights in the previous review as well as additional issues open a variety of perspectives to consider in examining various dimensions of the HGP. This section considers these issues in order to point to ongoing concerns in the HGP as a whole.

Context of Radical Individualism, Sociobiology, and Fascination with Science

Of critical importance is the context in which such issues are addressed. Several agenda items emerge. One is autonomy, the all-American virtue that is the source of much creativity in the culture and the driving value behind many of the positive reforms in medical ethics, such as a strong emphasis on informed consent. Yet given the way autonomy has been played out in our culture, the net result has been atomistic individualism. Moral actors are essentially self-actualizing and idiosyncratic monads whose interests, values, and goods cannot be known unless the window is opened. Consequently, no one can speak on behalf of the individual except the individual himself or herself. Such a narrow individualism has made it extremely difficult to discuss—let alone suggest—a range of common goods or common interests. Because of the primacy of the individual, the concept and reality of community have been harmed. People have few resources beyond themselves or specific contractual relations to which to turn in time of need. At a time when the understanding of the individual is to undergo a critical sea change because of the new understanding of the genetic structure and the role of genes, there are few, if any, communal resources to which to turn to address the question of identity. The available model—radical individualism—is showing its inherent weaknesses and limits at precisely the moment when a robust concept of the common good is needed. For if we do not look beyond ourselves to the society in which we live and recognize its profound effects on us, we will stand naked to our environment—and most likely suffer the fate of other hybrids.

Second, the developments in genetics are coming at a time when there is a renewed interest—through sociobiology—of understanding the theory of evolution biologically and socially from the Spencerian perspective of the survival of the fittest. This has led to the social celebration of the more predatory notions of human nature, which are reinforced by the dominance of autonomy. Asimov's three laws of robotics and the revisionist *Terminator 2* notwithstanding, sci-fi literature and films have focused on issues of dominance and the creation of an underclass through genetic engineering. Seldom does one read of efforts to genetically engineer or build kind, loving, wise, gentle people. A critical question emerging here is whether more of the same is needed.

Third, though there have been setbacks given the widespread publicity about data faking and other questionable ethical practices in science, Daniel

Callahan's decades-old comment about genetic engineering is still correct: "The scientific community and the general public are more than ever prepared to go ahead with it."[170] We are still fascinated by science, are enamored of its accomplishments and their effect on our lifestyle, and see biotechnology as a potential economic giant. Gains are being made in somatic cell therapy through genetic engineering, its impact on both agriculture and animal husbandry are proving significant, its contributions for medicine are being more clearly seen, and research is beginning. Although regulatory mechanisms are in place as a consequence of the earlier rDNA debates and the work of various presidential commissions, the dominant framework for analysis still seems to be the one earlier identified by Callahan: a right to seek what one wants if it does not harm others, risk-benefit analysis, and the cultural sense that it is better to try to do good than to prevent harm.[171] Yet precisely this matrix put us where we are today because this framework has great difficulty seeing beyond its narrow boundaries. If I am correct in my assessment that the first of the trio is the dominant one, then all else is judged by its impact on the individual. This gives little room to discuss and evaluate, much less give equal status to, the impact on society or social values.

I conclude that given this context, what is called for primarily is consideration of the social dimension of the person, the cultural environment that nurtures the person, and the common good.

Specific Issues and Directions to Take

STEWARDSHIP. In all discussion of genetic engineering the concept of stewardship plays a central role. Based in the creation story, the humans are stewards in that they are to conserve the garden in which they are placed. The basic metaphor connotes a sense of maintaining and continuing the status quo. Built into the stewardship concept is a notion of limits. And this sense is reinforced by the natural law tradition that defines morality as conformity to the laws of nature, which are the created manifestations of the eternal laws of God.

Two comments are appropriate here. First is the observation of O'Rourke and Ashley that such a concept is based on an understanding of God that is more Greek than biblical. Attempts to move beyond the garden, so to speak, are seen as Promethean strikes against the omnipotence of God. Another variation on this is to see attempts to improve the human as an insult to the intelligence of God.

Second, we have yet to thoroughly work through the concept of stewardship in the light of evolution. Given that the world—indeed the cosmos—is evolving and exists in a twofold contingency—it need not exist at all and it need not exist in this particular way—stewardship as a conservative concept seems totally unnatural.

Much more fruitful is the suggestion by O'Rourke and Ashley that God is a generous creator and that we are called to share in God's creative power. This echoes an earlier suggestion of Robert Francoeur that the better reading of steward is that of co-creator.[172] In this reading, to be a responsible steward is to participate in the ongoing adventure of evolution through seeking to unleash the potential within nature, including human nature. To be a steward is to be an actor, a designer, a planner, a co-creator.

Such a reading is a heady one, especially given Teilhard's understanding of the person as evolution become conscious of itself. The line between being co-creator and a creator is a fine one as numerous episodes of human history have shown. Yet a solid Christian reading of its tradition, informed by contemporary evolutionary theory, can draw no other conclusion than that the role of the person is to be an active participant in the ongoing drama of creation.

What is critical from an ethical perspective is the context in which this is understood and the uses to which this capacity is put. The exercise of stewardship also requires a continuous examination of the ends to which the legitimate powers of the person are applied and the reasons for which this is done. While hubris will continually remain a possibility and original sin and its ongoing effects a reality, there is no a priori justification for the assumption that we can neither transcend our own interests and look to the good of others and our environment nor implement a social and political structure to achieve such goods.

HUMAN NATURE: THEMATIC ISSUES.   One of the more enduring and recalcitrant issues is the nature of human nature. Books, journals, and databases are filled with competing models and claims about human nature. In this section I want to make two claims.

First, continuing my emphasis on the fact of evolution, we cannot claim that human nature is static or fixed. What our nature is now is not what it was nor is it necessarily what it will be. Given the fact of evolution, we cannot claim a human nature that was created distinctly and apart from the animal kingdom.

To say that does not mean that one cannot speak of human nature. Rather, what it means is that we must recognize that our claims are provisional and are time bound. In the past such claims may have seemed less provisional because evolution typically proceeded slowly. Now, however, given our knowledge of genetics and our increasing capacities, there is the distinct possibility for a speedup in the rate and direction of evolution. The map of the human genome will open vast possibilities before us, and though the rate of actually achieving specific changes will typically be slower than the development of our aspirations and plans, nonetheless the genetic map offers a dramatic possibility for directing evolution.

A second theme is whether human nature is the sum of its parts or whether there is a transcendent dimension to the human self. This issue is a key one, for on it hinges the validity of various strategies for intervention in human nature. If one assumes, for example, that human nature is the sum of its parts only, then one will have to argue that all human activity at root is genetically caused. The implication is that human nature and human actions and capacities can be changed exclusively or primarily through genetic interventions. For a change in human nature to occur under our own direction, all that is required is more knowledge, more skill, and more time. But eventually, the genome will reveal its secrets, and we will be able to make purposeful and directed changes in human nature as a consequence of altering the genetic structure of the individual.

If, however, human nature is more than the sum of its parts, genetic knowledge will continue to be important, but it will not hold a place of primacy. Other dimensions—the environment, culture, and the person as a self—will have to be included in any strategy for change. In this model, although change is possible and a good, specific changes will be more difficult to achieve and less targetable.

HUMAN NATURE: COMMENTARY. While the Christian tradition asserts strongly and clearly that human nature is more than the sum of its parts and indeed holds a dualism that has been expressed in various ways—animated flesh, incarnated spirit, substantial unity of body and soul—let me make a suggestion that points to such a reality and may help to ground the reality of transcendence without mandating a particular model of the relation.

I argue that there are experiences of self-transcendence, experiences in which we go beyond ourselves. One such experience is that of love, in which we are attracted to the beauty of another and seek the good of that individual for his or her own sake. In love we experience the being of another and are drawn to that value or goodness precisely because of its own value. To be sure, in this process we are transformed, changed, enhanced. But these are add-ons; they are wonderful experiences, but not the reason why we love or seek the good of the other. We love to revel in the other for his or her own sake. We celebrate precisely the otherness and its beauty and goodness as a center of value in and of itself.

Another experience is performing an action precisely because it is the right action to perform. One acts ethically because of the inherent worth and goodness of the act in question. Again, certain gains come to us from acting in this fashion, and probably more often than not, we do the right thing out of habit, out of lack of opportunity to do anything else, or out of fear. But we can and we occasionally do perform acts simply because they are the morally right ones. I

am not making an argument for deontological ethics here. Rather, I am pointing to the experience of doing the good for its own sake, of transcending self-
interest, of transcending the expedient and adhering to the good.

A final experience is that of freedom. Again the issue is that of going one way
rather than another, though the other was open to us. Negatively, the experience is one of not being compelled to act in a particular way. It is the source of
regret at an option not taken. Positively, freedom is the sense of self-actualiza-
tion that follows commitment to and actualization of a goal, it is the satisfac-
tion of knowing that an action is mine, and it is the ground of responsibility.

I argue that these experiences are critical, for they point to a transcendence
within human nature that resists reduction to a mechanistic explanation or to
one that looks at the parts only. If this is correct, then one has to attend to this
reality in planning genetic interventions. Häring, for example, sees the protection of freedom as an important criterion by which to evaluate genetic interventions; freedom distinguishes us from the rest of nature and is at the core of
our nature. Rahner also shares in this perspective by understanding humans as
freedom events, as self-creating beings. Such experiences open the possibility
not only of choice but also of choosing because of the value or goodness of the
object of choice. That is, choice reflects not only a capacity to choose but also
an act of valuing, a sizing up of the options, an implicit prioritization. Thus,
while freedom as the capacity to choose is significant, freedom as the capacity to
evaluate or appreciate is more critical.

Such capacities reveal a dimension of human nature that I take to be of enduring worth. Rather than function as a program or by instinct, we experience
a radical freedom at the core of our nature. With Häring, I argue that this is a
dimension of life to be protected; but I also argue that this fundamental capacity can lead to a more profound evolution of human nature. Freedom personal-
izes evolution and reveals its hidden possibilities.

CONSTRAINTS. Having affirmed evolution, a developing human nature, and
freedom, now I ask: Can limits be reasonably suggested? Is there a way to
ground an evaluation of genetic interventions?

One suggestion is provided by Rahner when he suggests that while we are to
become what we can be, we have not called ourselves into existence, and we do
not work with material or realities of our own making or creation. At the core
of our nature is a fundamental givenness of ourselves as created, as finite. This
grounds the reality of our being co-creators, not creators. That is, we do not
have final and ultimate disposal over creation, over evolution, over history. A
dimension of stewardship is the acceptance of finitude and the recognition
that, though free and the creation of a generous God, I am at liberty neither to
defile nature nor to confine it narrowly to my image and likeness.

Häring affirms the need to look at the history of sin and to remember what happens when the values of freedom and dignity are ignored. In particular, I would look to the history of the U.S. eugenics movement as a source for values and actions to be avoided. For that history reveals a very dark side of our society with respect to the treatment of both groups and individuals. An even more specific example is the point raised by O'Rourke and Ashley when they suggest that prenatal diagnosis may be leading us to a situation in which children are loved not because of who they are but because they manifest their parents' preferences.

Finally, Häring, C. S. Lewis, and Alasdair MacIntyre note that control is a critical ethical issue here. Häring's argument is that control is the contrary of freedom, and in seeking total control we seek to destroy freedom and thereby destroy human nature. Lewis observes that "what we call Man's power over Nature turns out to be a power exercised by some men over other men with Nature as its instrument."[173] And MacIntyre, after describing qualities that would be desirable in designing one's descendants, argues that an individual with these qualities would not engage in such a project.[174]

Control constricts the evolutionary process rather than opens it. Control assumes a privileged vision shared by only a few others. Control suggests a normative vision of the human. And in all of these suggestions there is a rejection of the givenness and finitude of our nature, as well as a rejection of the very openness of evolution itself. Seizing control represents an attempt to freeze development by making only one phase normative. Control in all its forms contradicts evolution and destroys freedom. Genetic interventions that seek control are problematic at best and need careful and extensive review.

Conclusion

The preceding literature review and commentary point to many concerns and issues in genetic engineering and the HGP. Many of these concerns are general, but specific issues have been raised by all authors. It is important, I think, to separate the concerns from either the view of human nature or the moral theory that supports the concern. There is, for instance, an overarching wisdom in the Catholic tradition that argues for the dignity of the person, a suspicion about power and control, and a recognition that physical interventions—whether medical or social—touch a person and not only his or her body. That wisdom stands independently of any of the specific claims and arguments of the magisterium or any theologians. Such wisdom needs attending to.

The most critical challenge comes from the attempt to incorporate an evolutionary perspective into ethics. Of particular importance is the grounding of ethical perspectives in an evolving world. Additionally, we need an ethic that both appreciates and critiques cross-cultural perspectives. Finally, we need to appreciate both the reality and the fragility of the gift of freedom. For it is our capacity to appreciate the good that will lead us to the openness and creativity needed to open evolution to the authentic future grounded in the gracious act of creation.

Notes

1. David F. Kelly, *The Emergence of Roman Catholic Medical Ethics in North America* (New York: Edwin Mellen Press, 1979), 244.

2. Ibid.

3. Ibid., 231.

4. Pius XII, "Christian Principles and the Medical Profession," in *The Human Body*, ed. the Monks of Solesmes (St. Paul Editions, 1960), 55.

5. Karol Wojtyla, *Love and Responsibility*, trans. H. T. Willetts (New York: Farrar, Straus and Giroux, 1981), 246, italics in original.

6. Pius XII, "Allocution to Midwives," *The Human Body*, October 29, 1951, 10.

7. Pius XII, "Allocution to the Italian Medical-Biological Union of St. Luke," *The Human Body*, November 12, 1944, 54.

8. Pius XII, "Allocution to the First International Congress of Histopathology," *The Human Body*, September 7, 1953, 199.

9. "Allocution to Those Attending the 'Primum Symposium Geneticae Medicae,'" ibid., 260.

10. *Gaudium et Spes*, in *Catholic Social Thought: The Documentary Heritage*, ed. David J. O'Brien and Thomas A. Shannon (Maryknoll, N.Y.: Orbis Books, 1982), 185.

11. Ibid., 186.

12. Ibid.

13. Ibid.

14. Ibid., 187.

15. Ibid.

16. Ibid., 200.

17. Ibid.

18. Ibid.

19. Pope Paul VI, *Humanae Vitae*, no. 9, quoted from *The Birth Control Debate*, ed. Robert G. Hoyt (Kansas City: National Catholic Reporter, 1968), 121.

20. Ibid.

21. *Humanae Vitae*, n. 12, 123.

22. Ibid.

23. *Humanae Vitae*, n. 13, 124.

24. *Humanae Vitae*, n. 17, 128.

25. Wojtyla, *Love and Responsibility*, 246.

26. Ibid., italics in original.

27. Ibid., 247.

28. Ibid., 246.

29. Ibid., 56.

30. Ibid., 56–57.

31. Ibid., 25.

32. Ibid., 56.

33. Ibid., 51.

34. Ibid., 66.

35. *Familiaris Consortio*, n. 32 (New York: St. Paul Editions, 1981), 51.

36. Ibid., no. 30, 49.

37. Pope John Paul II, "Biological Experimentation," *The Pope Speaks*, October 23, 1982.

38. *Donum Vitae*, n. 3, in Thomas A. Shannon and Lisa S. Cahill, *Religion and Artificial Reproduction* (New York: Crossroad, 1988), 144.

39. Ibid., 147.

40. Ibid.

41. Ibid., n. 3, 144.

42. Ibid., n. I, 4, 161.

43. Ibid., n. I, 4, 163, italics in original.

44. Ibid.

45. Ibid., I. 2, 149.

46. Ibid., 2, 150.

47. Ibid., 3, 151, italics in original.

48. Ibid., 4, 153, italics in the original.

49. Ibid.

50. Ibid., 6, 154.

51. *Medicina e Morale*, Gennaio-Febbraio 1990, 149–58.

52. Ibid., 151.

53. Ibid.

54. Ibid., 152.

55. Ibid., 153.

56. Ibid., 154.

57. Ibid., 158.

58. Catholic Health Association of the United States, *Human Genetics: Ethical Issues in Genetic Testing, Counseling, and Therapy* (St. Louis: Catholic Health Association of the United States, 1990), xiii.

59. Ibid., 11–14.

60. Ibid., 20.

61. Ibid., 25.

62. Ibid., 33.

63. Ibid., 35.

64. Ibid., 36.

65. Ibid.

66. Ibid.

67. Ibid., 37.

68. Ibid., 31.

69. Karl Rahner, "Experiment: Man," *Theological Digest*, February 1968, 61.

70. Ibid., 62.

71. Ibid.

72. Ibid., 67.

73. Ibid., 65.

74. Ibid., 63.

75. Karl Rahner, "The Problem of Genetic Manipulation," in *Theological Investigations* IX (New York: Seabury Press, 1976), 230.

76. Ibid., 243.

77. Ibid.

78. Ibid., 244, italics in original.

79. Ibid.

80. Ibid., 245.

81. Ibid., italics in original.

82. Ibid., 250.

83. Ibid., 236, italics in original.

84. Ibid.

85. Ibid.

86. Ibid., 246.

87. Ibid., 248, italics in original.

88. Ibid., 248–49.

89. Ibid., 249.

90. Bernard Häring, *The Ethics of Manipulation* (New York: Seabury Press, 1975), 4.

91. Ibid., 11.

92. Ibid., 17.

93. Ibid., 28.

94. Ibid., 39.

95. Ibid., 50.

96. Ibid., 51.

97. Ibid., 52.

98. Ibid.

99. Ibid., 57.

100. Ibid., 60.

101. Ibid., 70.

102. Ibid.

103. Ibid., 71.

104. Ibid., 72.

105. Ibid., 76.

106. Ibid., 161.

107. Ibid., 163.

108. Ibid., 170.

109. Ibid., 183.

110. Ibid., 184.

111. Ibid., 185.

112. Ibid.

113. Ibid., 211.

114. Richard A. McCormick, *Health and Medicine in the Catholic Tradition* (New York: Crossroad, 1984), 51ff.

115. Ibid., 53, italics in original.

116. Ibid., 54, italics in original.

117. Ibid.

118. Ibid., 57.

119. Ibid., 58.

120. Ibid., 141.

121. Richard A. McCormick, "The Ethics of Reproductive Technology," in *The Critical Calling: Reflections on Moral Dilemmas Since Vatican II* (Washington, D.C.: Georgetown University Press, 1989), 344.

122. Richard A. McCormick, "Genetic Technology and Our Common Future," in *The Critical Calling*, 266.

123. Ibid., italics in original.

124. Ibid., 267.

125. Ibid.

126. Ibid., 268.

127. McCormick, "Reproductive Technology," 345.

128. Charles E. Curran, "A Methodological Overview of Fundamental Moral Theology," in *Moral Theology: A Continuing Quest* (Notre Dame, Ind.: University of Notre Dame Press, 1982), 38.

129. Ibid., 42.

130. Ibid., 44.

131. Ibid., 47–48.

132. Ibid., 50.

133. Ibid., 51.

134. Ibid.

135. Charles E. Curran, "Genetics and the Human Future," in *Issues in Sexual and Medical Ethics* (Notre Dame, Ind.: University of Notre Dame Press, 1987), 112.

136. Ibid., 117.

137. Ibid., 119.

138. Ibid., 124.

139. Ibid., 127.

140. Ibid., 129.

141. Charles E. Curran, "In Vitro Fertilization and Embryo Transfer," in *Moral Theology*, 124.

142. Ibid., 125.

143. Ibid., 131.

144. Ibid., 130.

145. Ibid.

146. Benedict Ashley and Kevin O'Rourke, *Health Care Ethics: A Theological Analysis*, 2d ed. (St. Louis: Catholic Health Association of the United States, 1982), 306.

147. Ibid.

148. Ibid.

149. Ibid.

150. Ibid., 307.

151. Ibid.

152. Ibid.

153. Ibid., 308.

154. Ibid., 324.

155. Ibid., 325.

156. Ibid.

157. Ibid.

158. Ibid.

159. Ibid.

160. Ibid., 326.

161. Ibid.

162. Ibid.

163. Ibid., 327.

164. Ibid.

165. William E. May, "Biomedical Technologies and Ethics," *Chicago Studies* 11 (1972): 245–49.

166. Ibid., 254.

167. Ibid., 255.

168. Ibid., 256.

169. Ibid.

170. Daniel Callahan, "The Moral Career of Genetic Engineering," *Hastings Center Report* 9 (1979): 9.

171. Ibid.

172. Robert Francoeur, "We Can—We Must: Theological Reflections on the Technological Imperative," *Theological Studies* 33 (September 1972): 429.

173. C. S. Lewis, *The Abolition of Man* (New York: Macmillan, 1957), 34.

174. Alasdair MacIntyre, "Seven Traits for the Future," *Hastings Center Report* 9 (February 1979): 5ff.

[8]

Mapping the Normal Human Self: The Jew and the Mark of Otherness

Laurie Zoloth-Dorfman

THE LOCUS OF INQUIRY in Jewish medical ethics has focused largely on the topics most common to the tangible clinical encounter. This chapter discusses a subject that is at one remove from this encounter and calls for reflection on the Human Genome Project (HGP)as it addresses the biological basis of human "nature."[1] The HGP is a federally funded research project constructed as a multicentered inquiry into the map of the human genome. It has been lavishly described as "the Holy Grail of Biology,"[2] as explaining "the key to what makes us human, what defines our possibilities and limits as members of the species *Homo Sapiens*."[3] "The knowledge will undoubtedly revolutionize understanding of human development, including the development of normal characteristics, such as organ function, and abnormal ones, such as disease. It will transform our capacities to predict what we may become and, ultimately, it may enable us to enhance or prevent our genetic fates, medically or otherwise."[4]

The project raises disturbing questions both in the philosophical construction of the enterprise and in the possible practical applications of the data obtained. That the project will be completed is a certainty. That the application of the information will reconstruct our understanding of what is "diseased" and what is "normal" is equally apparent. It is largely a project about power and truth as much as biochemistry. But the normative power assumed by such an enterprise is itself problematic. It is the contention of this chapter that even the extraordinary and redemptive acts of science must occur in the context of a discursive and democratic community that will bear the responsibility and the meaning of that science. Such a community must be aware of a shared history and of a mutuality of responsibility that such history insists upon.

For this reason Jewish medical ethics[5] must bear a particular obligation in response to this philosophical inquiry. For Jews, the denotation of the "normal" and of the "truth" of scientific categorization of "normal" and "disease" has historically been problematic. Ideas about deviance, the animal, and the otherness of Jews have historically been linked with systematic exclusion and oppression of the Jews. The arena of scientific inquiry has been no exception to this; rather, the sciences of eugenics, physiology, and medicine have been utilized to describe the Jew as first excluded, then dangerous to the health of the normal public welfare.

This chapter examines two issues. First, I argue that the search for the perimeters of the normal presents particular problems in a post-Shoah[6] context and that the historicity and contextuality of the Human Genome Project matter greatly. In this context I make reference to the recent literature on the issue of the body and the social construction of the Jew's body to reflect the problems of the construction of a normal self. Further, I argue that as a result of such construction of the normal, serious issues of justice arise in the application of the project, and that the implications of the justice issue may be differently experienced not only by Jews, but also by others defined and marked as physically deviant—women, persons of color, and people with disabilities.

Second, I examine selected textual sources that describe and regulate *halakhic* responses to scientific inquiry. Modern Jewish ethical response to the HGP has been enthusiastic, praising the intellectual enterprise itself and hailing the possible therapeutic benefits.[7] Earlier texts on the nature of human creation present a more nuanced view. An examination of such texts leads to another view of selected Jewish attitudes on the problem of suffering, on the meaning of illness, and the issues of limits in medicine.

The Measure of the Normal

New knowledge of the human genome will yield new powers to transform the human reality. The fascination with the limits of the normal self and the quantitative measurement of this self are not new. The new instrumentation of the eighteenth century only accelerated a process of caliperization and categorization that focused on the knowing of the world by the most accurate observation and tabulation of it. The marked body of the Jew has long been a source of reference against which to measure the normal human body, read generally as the body of the Christian male.[8] Jews were different, "marked in the flesh" by the act of circumcision. Since the marker that was evident on the genitalia was hidden from plain and public sight, external signifiers became the subject of deferred scientific inquiry in the nineteenth century. Sander Gilman, in a groundbreaking book about the historical view of the Jew's body, reads this history carefully in light of the science that marginalizes and pathologizes the "different" Jew.

> The very analysis of the nature of the Jewish body, in the broader culture, or within the culture of medicine, has always been linked to establishing the difference (and dangerousness) of the Jew. This scientific vision of parallel and unequal "races" is part of the polygenetic argument about the definition of race within the scientific culture of the eighteenth century. In the nineteenth century it is more strongly linked to the ideas that some "races" are inherently weaker, "degenerate," more at risk for certain types of diseases than others. In the world of nineteenth century medicine, this difference becomes labeled as the "pathological" or "pathogenic" qualities of the Jewish body.[9]

The mapping of the "normal nose" (the clearest stand-in for the troubling issue of the different penis) and the mapping of the "normal foot" were earlier examples of such a quantitative search, both constructed to frame the debate about whether the "blood of the Jew is the same as ours."[10]

As early as the 1200s the mythology of the blood libel against the Jew included speculation in Christian texts that in fact, Jews inhabited not real human bodies, but grotesque and demonic forms. The point of the search for a Christian innocent to murder was ascribed to the need to ingest the blood of the Christian, because only with the blood of the Christian in his veins could the Jew assume a human form.[11] The reference to and the fascination with what blood ran in the Jewish veins are linked to this persistent association.

By the nineteenth century, the sign of difference was more highly nuanced and biased not in lurid speculation and religious fervor, but in clinical research science, as science began to challenge religion as the dominant ideology with which to explain phenomena and rank order the world. The foot of the Jew was "discovered by the medical scientist of the nineteenth century to be flatter than that of the gentile" (this observation based on examinations by army doctors in the German Empire). Based on this "fact" an entire set of second assumptions arose about the nature and the meaning of a Jew's ability to be a soldier (read a man) and hence a citizen.

> In the nineteenth century, the relationship between the image of the Jew and that of the hidden devil (and the cloven hoof of the devil) is to be found not in a reli-

gious context but in a secularized scientific context . . . the pathogenic foot of the "bad" citizen of the new national state. The political significance of the Jew's foot within the world of nineteenth century medicine is thus closely related to the idea of the "foot" soldier, of the popular militia, which was the hallmark of all of the liberal movements of the mid-century. . . . Jews could not be good soldiers due to their weak constitution, and its public sign, "weak feet."[12]

Not only the flat-footedness of the Jew made him unacceptable, and in fact a danger to fellow soldiers, but the Jewish gait could be mapped and charted, then contrasted with the gaits of Aryans. There are actually detailed gait charts that "scientifically" and carefully note how the "normal gait" differs from the Jewish, the criminal, and the epileptic gait.[13] And an abnormal gait is not a neutral sign of disease. It is the mark of the last stages of the syphilitic, the disease long linked to the Jew, who was accused of spreading the disease in Europe, becoming marked by the disease, and of having a deviant sexuality again associated with a "deviant" penis. Hence the moral meaning of the different gait was that it revealed the hidden Jew, and marked the Jew as different, despite the other signs of regularity.

Jewish scientists of that era uncritically accepted the basic premise, challenging neither the "facts" as reported as biased nor the premise that arched feet were, in fact, normal or a significant variation with some behavior consequences.[14] Their response was to agree that the Jewish foot was deviant, not normal, and was problematic, attributed to social rather than genetic factors. The Jew, with proper exercise and relief from the trauma and distortions of urban life in the ghetto, could perhaps achieve normalcy and productivity. It was not the racial fault or fate of the Jew to be different in this way, argued Jewish scientists, but a correctable social issue.[15] Hence, the notion of fault and exculpation has long been an issue in the identification of difference. Here note the reverse in the modern understanding: if a genetic and physically "real" marker can be found (such as the "gay gene"), then perhaps the fault for the deviance can be shifted from the social to the medical. Thus, the individual is not to blame and is treatable via medical reorganization. This is the central metaphor of the entire HGP: that disease (read deviance defined as problematic to the medical culture) can be eradicated by neutral medical means. Our fate is not in our stars.

Gait mapping and foot-arch measuring charts were not the only such endeavor. Nose charts, drawn in the same period, described the normal Aryan nose and the distorted, not normal, Jewish nose.

Jews bear the sign of the Black, "the African character of the Jew, his muzzle-shaped mouth and face removing him from certain other races. . . ," as Robert

Knox noted at mid-century. The physiognomy of the Jew which is like that of the Black: ". . . the contour is convex; the eyes long and fine, the outer angles running towards the temples; the brow and the nose apt to form a single convex line; the nose comparatively narrow at the base, the eyes consequently approaching each other; the lips very full, mouth projecting, chin small, and the whole physiognomy, when swarthy, as it often is, has an African look."[16]

And the nose was not merely different; it was deviant and diseased. Many Jews introjected the standard image of normalcy and thought of themselves as physically defective. Aware of their "deformity," many became terribly depressed, melancholic. For such disease, there was a cure: the nose job. This procedure, like the admonition to Jews to play sports to make their feet normal, was based on the acceptance of medical accounts of the construction of disease and the medical solution, in this case, surgery to create a nose that would be fitting and not betray the dangerous Jewishness overtly. The "facial-plasty" was invented by a Jewish surgeon, Jacques Joseph, who opened a practice in Berlin in 1898 and operated on "anyone suffering from a Jewish nose," even for nothing, until 1933 when the surgery to render the increasingly visible, yellow star-wearing, and "diseased" Jew invisible was forbidden.

That the physical characteristics of the Jew and the psychological ones were linked was clear. That was easy to understand based in the racial theories of the German scientist that were widely believed by the Jewish theorist. Sander Gilman notes that even Maurice Fishburg, while defending some proportion of the population against the diagnosis of endemic hysteria (linked not only to women, but to Jews), quotes the work of Richard Andre:

No other race but the Jews can be traced with such certainty backward for thousands of years, and no other race displays such a constancy of form, none resisted to such an extent the effects of time as the Jews. Even when he adopts the language, dress, habits, and customs of the people among whom he lives, he still remains everywhere the same. All he adopts is but a cloak, under which the eternal Hebrew survives; he is the same in his facial features, in the structure of his body, his temperament, his character.[17]

This link to hysteria was not only, like with women, the recognition of powerlessness and place in a repressive society; it was linked with the excessive sexuality of the Jew (as the etiology of the female hysteric is sought in unbridled obsessive sexuality),[18] and this also was linked to the inbreeding of the "incestuous" Jewish community. As Gilman further notes, there is a persistent link between dangerous sexuality, genetics, and disease in the regard of the Jew.

Finally, the skin of the Jew, both the color (swarthy or black) and the complexion (seen as diseased and poxed), was said to be distinct and in fact ugly. It was not a mere aesthetic difference. The "black" color and pocked complexion, like the other characteristics, were clinically and scientifically linked to the sexual mark, the symptomatology of the syphilitic, as noted above a disease long linked and blamed on Jews in Europe. The disease that blackened the skin was the disease that the Jew carried into the pure, clear, Christian body and into the body politic. This outward evidence of the Jewish difference could be altered, and anecdotal evidence reflects that, as the Jew emerged from the poverty of the earlier nineteenth century, the skin color lightened, making Jews look more like the citizens of whatever country they lived in. But the essential Jewish demeanor was not diluted, not hidden by dress or manner. Long into the twentieth century the mark of the race, the mark of the circumcision, and the "singular piercing gaze" marked the Jew as outsider, as other than normal.

The linking of the not normal type with disease and danger has been repeated, and not just in the nineteenth century, but whenever the claim could be made for the difference of the Jew. By the twentieth century, of course, the scientific texts that had trained generations of physicians about the normal, and exactly who was and was not diseased, were transmogrified into the political texts of the Nazi regime.[19] The neutral and demonstrable physicality, accepted uncritically even by Jewish scientists, stood in place, to be applied to horrifically technical "solutions," in turn administered by many German medical doctors who had come to both construct and believe deeply in the facts as they saw them with their own eyes.

It was not only the German experience that ought to trouble the discourse, or the excesses of unbridled totalitarianism that ought to disturb the debate. The United States, the HGP defenders argue, is a deeply democratic country with long-standing traditions of diversity, unlike the cultivated hatreds of Europe.[20] Hence, more disturbing is the American application of early work in genetics. Sociologist Troy Duster, in a critical study of the HGP, has noted the extraordinary power of genetics in the construction of social policy that is designed to marginalize and exclude the visibly different. Duster has stated: "Empirical research on the genetics/IQ controversy and the attendant policy injunctions have had a remarkably varied history. However, one consistent pattern emerges: the more privileged strata have at each juncture raised the 'genetic' question about those at the lower end of the socioeconomic ladder."[21]

To illustrate, Duster used the example of the Jews. At the turn of the century, Jews from Eastern Europe and Russia immigrated in large numbers to the United States, and nativist fears arose that the influx of immigrants would pollute the real American gene pool. The fears were specifically phrased in genetic terms. And not only were Jews immigrating in record numbers; they tended to do quite well in school, emerging as competitively superior to Gentiles relative to college admissions. Yet on standardized testing, introduced in the early part of the century, Jews tended to do rather poorly. In fact: "In congressional testimony to justify new legislation to stop the flow of this 'lower form of human life,' the scientific IQ test becomes a powerful justification. Jews (and Russians and Italians) had been revealed by the Binet test as genetically inferior, and more prone, biologically, to 'feeblemindedness.'"[22]

In 1924, restrictive immigration laws were passed in the United States, a social policy in large measure buttressed by the rationally constructed measure of disease prevention: more than 40 percent of the Jewish immigrants showed up as feebleminded on such tests.[23] However, by the 1940s, Jews tended to perform with consistent superiority on Binet tests. Duster also reveals that when Jewish test scores could no longer be counted on to exclude Jews from Ivy League colleges, quotas on Jewish students were imposed. These quotas persisted into the 1970s.

From "Image" to Eugenics

The Human Genome Project is highly sensitized to the problem of its links with past eugenics movements. Nearly every article takes great pains to distance itself from this history.[24] It is a truism of genetics that since one is aware of this connection, it can be avoided. But the problem is far deeper. The eugenics texts that intellectually supported the Shoah not only justified that particular horror, but were in and of themselves formulative in their ideation. Perhaps the most disturbing aspect is the way that the Jewish physicians and researchers were eager to do the medical work that would allow the appearance of normality to mask the reality of significant racial difference: how willing they were to cut off parts of their bodies to make them fit the map. This problem is no mere historical anachronism. The seeking of the perfect (read Aryan body) persists, reflected in the everyday American belief that image is everything. Conforming to the image is a consuming enterprise. Every week in my Sunday paper, in the aptly named *Image* magazine, the following advertisement appears:

> *A Guide for Cosmetic Enhancements: a More BEAUTIFUL YOU*
> *Nose Enhancements*
> *Daniel B. Cohen, MD*

A "before" picture is shown of a young, rather swarthy Jewish-looking woman with long dark hair: she looks Jewish because of her large hooked nose. She is wearing three earrings, gypsylike in her ear. The "after" picture is remarkable: her nose is upturned and shortened, her skin is lightened, and her hair is cut and bobbed as well. She wears one set of small gold earrings. "After one week," reads the ad, "the cast is removed. There will be some swelling, but the new nose becomes more beautiful and refined over time."[25] There is also a picture of Dr. Cohen,[26] smiling in his white coat, over a brief list of his scientific credentials. He is board certified, we are assured, in the specialty developed by Jacques Joseph nearly a century ago. The ad notes that he will also diminish the size of the breasts, lips, and hips of the body—large breasts and hips being similarly associated with Jewish women, specifically the overly large, overly consuming Jewish mother of the popular imagination. All the parts of the offending Jewish woman's body can be literally cut off to fit her into a smaller space in a Gentile world: it is strictly a medical procedure, a cure that the doctor offers the patient for the disease of looking too Jewish. Here we have the Jewish surgeon, the *Cohen* of the medical world, in fact, medicalizing the problem of the Jewish nose: the ugly and unrefined nose, which can be treated with surgery, a cast, and "mild medications."[27] Medicine of the present can "treat" the problem of the Jew's body with the tools of the armamentarium: surgical knives, medication, cosmetics. In the future, the tools may well be genetic manipulation,[28] and if the context for the yearning is unchanged, the call for the use of these tools to obliterate the mark of the Jew will be simply transferred to this arena.

If science was the measure of the real itself, then we are addressing a much larger problem than the danger of discrimination. When we speak of the HGP's classification scheme, we are trafficking in the market of truth criteria. What the science of genetics has historically done is to frame the questions in the language of the dominant culture, and the questions themselves, as Duster indicates, carry the power to define personhood or, rather, successful personhood. The claim to legitimacy, to the authority of place, was historically raised by the nineteenth-century science itself. Jews were not of the normal place of the nineteenth century. They were out of place geographically and out of place temporally, yearning for Jerusalem unconquered and pre-Christian, continuing to recite prayers of the first century, carrying with them the scent, the gaze, the gesture, that were of this lost Jerusalem, entirely their own.

As Daniel Boyarin maintains in his work,[29] the Jews were representative not only of religious difference, but also of physicality, of the body and of the carnal itself. The fixation of the European culture with this aspect of definitional naming of the other is persistent, and has as much to do with a cultural preoccupation with the containment of the other as the purity of the scientific inquiry.

"Is Jewish Blood the Same?"

In writing this chapter, I became interested in a particular possible finding of the HGP. Could it be true that the HGP will reveal some Jewish genotype? There are several reasons to suspect that this might happen. Tay-Sachs disease is long linked to the Ashkenazic Jewish populations. Canavan disease, another neurologically devastating and degenerative disease, is also linked to Jews nearly exclusively. In an article in *Natural History*, Jared Diamond, professor of evolutionary biology at UCLA, discussed the topic of the difference of the Jews as it is expressed in the Jewish karyotype.[30] Diamond asks the question of origins and persistence: Are the genes found arrayed in the Jew of today the same or different from the genes of the Jews that emerged into medieval Europe from the Diaspora of the first century? Are the blood of the Jew and the body of the Jew expressive of a particular genetic structure, or are they like the Europeans among whom they lived for centuries?

Diamond notes the obvious similarity of Jews to the peoples they live among. Gilman comments on this as well: it is the blond Californian child of the darker New York parents syndrome. The Jews from Yemen "look" like the Gentiles of Yemen, the Iranian Jews "look" Iranian, and the Ashkenazic Jews, especially from the most northern countries of origin, tend to be blonder and taller than the Sephardic Jews. Moreover, the initial genetic studies of single gene markers found in the 1940s, for instance, the deficiency of the enzyme G6PD is rare among Ashkenazic Jews and Eastern European Gentiles, but it is common among Mediterranean Jews and Gentiles. Blood types tend to be geographically specific. These results would tend to suggest that Jews had either from freely chosen or coerced (rape being a common part of conquest) sexual liaisons intermingled the gene pool. But Diamond observes that the genetic characteristics that are found in common with neighbor populations are exactly those that are based on the most highly adaptive mutations. Diamond claims:

> Among Darwin's most important insights was his concept of natural selection: depending on the environment, certain traits permit an animal to survive (and to pass on its genes) with higher probability than do other traits. Thus two populations derived from the same ancestral populations may diverge genetically in different environments, while populations from different ancestral stocks will tend to converge in the same environment. One of the most familiar human examples involves skin color: tropical peoples tend to genetically have darker skins than temperate zone people. . . . Conversely, modern Europeans speaking related languages and probably sharing a common genetic ancestry, exhibit a striking gradient in skin color from north to south.[31]

Similarly, the deficient G6PD gene tends to protect against malaria: needed near the Mediterranean, but not in, say, Poland. ABO blood groups as well are now thought to be distributed relative to the distribution of infectious disease, such as smallpox or cholera. Since microorganisms are more common in some climate and geographic zones than others, it is not surprising that blood groups are also specifically adaptable. In the United States, for example, the sickle-cell trait is far less common among African Americans than it is among Africans who live in areas where malaria continues to be a threat. This cannot simply be attributed to intermarriage, since many other genetic characteristics are identical, and since the same differential is also noted between Africans who live in areas with no malaria and those who live in highly malarial areas, who are in other ways genetically similar.[32]

That such shifts can happen rather quickly should not startle us: Diamond points to the wagtail, yellow in some countries, differently patterned in others, but always a wagtail. British moths have darkened since the Industrial Revolution, insects have developed DDT resistance, and new strains of flu evolve every few years. That the nineteenth-century scientists noted that the Jews were swarthy, and that by the twentieth century some Jews had perceptible changes in skin tone, was in part due to the emergence from the ghetto and the diseases of poverty, and in part due to nutritional advantages; but it was also in part due to this adaptive mechanism.

But interestingly enough, relatively few genetic mutations are of this adaptive type. Judging by neutral markers, the non-Jewish contribution to the Ashkenazic and Sephardic Jewish gene pool has been low.[33] More recently, more highly nuanced studies show:

> In their fingerprints, Rhesus blood group frequencies, haptoglobins, and several enzyme markers, Ashkenazic Jews resemble Sephardic and Yemenite Jews and differ from Eastern European Gentiles. Furthermore, in these respects Jews resemble many Gentile peoples of the eastern Mediterranean, such as Samaritans, Armenians, Egyptian Copts, and Syrian, Lebanese and Palestinian Arabs.[34]

If one actually made charts of the human genome, then they might look rather like the anthropological charts of the nineteenth century, with columns marked "Aryan," "Semitic," "African," and so on. At issue is not that diversity exists; it is, rather, that our culture has such trouble with diversity as a neutral concept. Is it normal for the human to be G6PD deficient? Is the carrier status a disease? Are persons with late onset symptoms that may or may never express "disabled"? Are they, in the words of Albert Jonson, bioethicist, "unpatients"?[35] Will the map of the normal be the map of the American (read European) karyotype in the way the gait of the normal was the gait of the European?

The Search for the Traditional Text

The second question this inquiry will address is a related point: How ought we frame the response to the HGP, given these serious concerns for its just application? As a Jewish scholar, with a commitment to addressing the issues in bioethics, how do I frame the Jewish response to such an endeavor?[36] To what extent does the current enthusiasm for the project by traditional Jewish authorities reflect the accurate voice of the textual tradition, and to what extent is it influenced by the same shaping forces that affected the Jewish scientists of the nineteenth century: Sigmund Freud and Wilhelm Fliess? This discussion is intrinsically linked to the first issue that I have lifted up—the historical avoidance of the inbred mark of the Jew, read as the disease of the Jew. The philosophical and normative ethical issues that surround the mapping of the normal are impossible to grasp without a historical view of such work; but the energy of the project is similarly incomprehensible without an appreciation of the potential for cure that is promised by gene mapping. This is the linking concept between the country of pure science and the country of frank eugenics, for the charting of the normal or the perfect is the naming of the oldest yearning—that to be a perfectly conceived human is to be free of suffering.

First a note on method. I am committed to the reflections on the problem of the HGP, as I am methodologically in all work I do in the field, to research that looks carefully at the primary texts of *halakhic* literature, the Talmudic sources, and the *responsa* literature. It is a cornerstone of this method to begin the reflection on the nature of the right act and its justification with a thorough search of the texts where analogous cases have been discussed and debated.

Let us review some of the normative responses. The general thrust of Jewish response to medical advance has been positive, even optimistic; linked to the notion that advanced scientific inquiry is a part of *tikkun olam*, the mandate to be an active partner in the world's repair and perfection. Judaism is not a nature-based religion; the very assertion of circumcision rests on the notion that the body is neither sacred nor immutable. The belief that the personal body is a property that belongs to the self alone is a late and nontraditional response to medical decision making. Traditionally, nearly all commandments could be abrogated to permit acts of lifesaving intervention or healing. Characteristically, "Judaism does not interfere with a physician's medical prerogative, providing his considerations are purely medical in character."[37] The permission and the obligation to heal come directly from the Torah text of Exodus and Deuteronomy, as interpreted by the Talmud: "The school of R. Ishmael taught *and heal he shall heal (Exodus 21:19)*. This is [the source] whence it can be derived that the authorization was granted [by God] to the physician to heal."[38] And further: "How do we know [that one must save his neighbor from] the loss of himself? From the verse *And thou shall restore him to himself (Deut. 22:2).*"[39] A positive attitude toward medicine stresses that the recourse to prayer and faith alone is incomplete without the full use of all of the resourcefulness of which humans are capable. This capability is a God-given gift, part of the work of stewardship to which persons are entasked in Genesis.[40] Ethicist Elliott Dorff cites another text that directs the general attitude toward the medical endeavor, and by extension the work of the HGP. The physician's work is legitimate and, in fact, obligatory. Rabbi Akiva and Rabbi Ishmael are walking in Jerusalem, and they encounter an ill person, who asks for their expertise in finding a cure. They tell him, but the man is puzzled: After all, are not the rabbis transgressing the will of God who made him sick in the first place by curing him? They answer by asking him about his work. He is a farmer who works in the vineyard, created by God: Does he not alter the world that God created by his work? The text continues as follows:

> [He answers to them,] "If I did not plow, sow, fertilize, and weed, nothing would sprout."
> Rabbi Akiva and Rabbi Ishmael said to him, "Foolish man. . . . Just as if one does not weed, fertilize, and plow, the trees will not produce fruit, and if fruit is not produced but is not watered or fertilized, it will not live but die, so with regard to the body. Drugs and medicaments are the fertilizer, and the physician is the tiller of soil."[41]

Dorff notes the mandate to be a partner in creation and generally acclaims genetic engineering "one of the wonders of modern medicine." While he recognizes the potential for eugenic uses, "the potential benefits to our life and our health are enormous," and hence, research ought to continue.[42]

Although no specific texts address the issues of the use of research science specifically, the Talmud is replete with stories about the general ability of the rabbis to examine closely the abortus itself or to observe closely specific medical conditions.[43] On the other hand, no *halakhic* texts forbid basic research either. David Bleich holds that these phenomena are characteristic of several modern problems in medicine, ones where there are no clear textual referents. In a book on Judaism and healing, he has used texts that refer to the necessity to build fortifications around cities. The community must build walls in the face of danger, but the obligation that the community has to protect itself against imminent danger does not extend to danger that exists in the not-yet-existing future. Hence, by extrapolation, genetics work that promises the very real chance of saving a life is an obligation to pursue even in the face of other theoretical dangers. In Bleich's view, the premise is clear.[44] The science as promised offers enormous potential benefits to horrific and fatal disease states. The HGP will lead to such advances. Further, scientists who do medical research ought to be assumed to be working for the welfare of humanity, not to its detriment.[45] Bleich notes that in terms of Jewish *responsa* literature, the possibility of "hard science" was relatively recent. No significant commentary emerges until the mid-eighteenth century, and there only vague reference to basic research is found.

Given such positive *halakhic* responses, the nearly universal communal response to all genetic advances that can promote health and increase fertility has been enthusiastically positive in the Jewish world. Note here the logical progression: the HGP simply describes and organizes the data that are before us, describing and shaping our view of the "factual" and the possible. Such data are seen *halakhically* as approving the next scientific breakthrough, and as such is supported. The HGP makes possible the rapid development of the next step of somatic research, identifying the gene for fatal incurable disease. Knowing the location of the abnormal gene makes possible its elimination (at this stage this is achieved by its avoidance). And the Jewish community has been quick to embrace this part of the process both communally and *halakhically*. Only three countries have more in vitro fertilization (IVF) centers than Israel (the United States, the United Kingdom, and Australia).[46]

The success story that nearly all commentators on the HGP point to, including Duster, is the rapid response of the Jewish community to the discovery of the single gene disorder of Tay-Sachs, a devastating neurological condition that is universally fatal. Millions of Jews in the United States and Israel were tested for carrier status and counseled on how to avoid giving birth to such children. Testing was carried out in a climate of urgency. Rabbis gave pulpit sermons about its importance. Major *halakhic* authorities ruled quickly on the use of the test and the permissibility of permitting abortion in these special circumstances. The number of children born with this condition in the Jewish community fell dramatically.

But now the technology has allowed the process to go much farther. In U.S. and Israeli Ultra-Orthodox circles, where it is possible to organize an arranged marriage, one of the increasingly accepted criteria is having a negative carrier status for Tay-Sachs, Gaucher's, and cystic fibrosis.[47] Eight thousand people were tested in 1993 alone, and sixty-seven marriages were "avoided." In the case of Gaucher's, a late onset lipid storage disorder, the symptoms generally appear after age forty-five, sometimes are identified only by autopsy, and can be entirely controlled with medication. In the case of cystic fibrosis, the disease can vary greatly in intensity, from severe respiratory conditions that need medication to the propensity to frequent colds. But if the concept of prenuptial and prenatal screening is *halakhically* acceptable for Tay-Sachs, and the technology exists to uncover more and more diseases, then the process shifts perilously close to the eugenic imperative. Disorders are increasingly being identified as genetically based. Articles in the popular secular and Jewish press set the tone of the discourse:

A decade from now, a lab technician will be able to draw a little blood and chart your whole genetic profile—what diseases you'll get, if you are naturally inclined to aggressiveness. . . . You'll be able to get a similar chart on your unborn fetus, including whether it is likely to be short or tall, dark or fair. You may even be able to order up a child with the characteristics you desire.[48]

A front-page article in the *New York Times* reports on the prenuptial counseling "that just mushroomed" after a testing service for Tay-Sachs expanded to include other tests as they became available. "So why leave this up to God, God has enough to do," says one of the project's founders. The marked and the diseased Jew can be avoided by others. The name of the project in the United States is *Dor Yeshorim*, Hebrew for the "generation of the righteous."[49] Further identification will reveal other genetic links understood as "abnormal." Immanuel Jakobowitz, one of the leading and well-respected *halakhic* scholars, touched off a controversy in London when he gratefully acknowledged the findings that homosexuality in men might be genetically based. "That means there is some hope that this tragedy can be cured," he is quoted as saying.[50] A child born with the gene could be "trained to help him overcome his predisposition." This view is consistent with views expressed earlier and is based on his interpretation of the biblical text that is the basis of the Jewish mandate to heal.[51]

The Search for the Alternative Story

In reflecting on this positivistic tradition, and in light of the profoundly troubling parallels found in the eighteenth- and nineteenth-century scientific texts relative to normalcy and the Jews, I wondered if other texts as yet unutilized outside the strict parameters of the *halakhah* could be found that questioned the philosophical base for the HGP in some meaningful way. What I was seeking was not a way to be contentious, but texts that looked not only at the consequences of the work as a way of assessing its normative value but also at issues of deontological concern.

The claim by the founders of the project that they are learning the secrets of human creation, their claim that they are close to "uncovering . . . the true knowledge" of how a human being is formed, and the claim of "knowing the elements of the building blocks of human life" so often cited by the HGP's supporters recalled a theme in Jewish sources. I turned to the Jewish *aggadic* tradition about the persistent efforts to actually create a human person—the Golem legends and text.[52] This theme recurs frequently in the tradition. According to Moshe Idel, its most influential mention is in the Talmud: "Rava said: If the righteous wished, they could create a world, for it is written, 'Your inequities have been a barrier between you and your God.' For Rava created a man and sent him to R. Zeira. The Rabbi spoke to him but he did not answer. Then he said: 'You are from the pietists: Return to dust.'"[53]

What is occurring here? Rava demonstrates that creation of some type of human life is possible: the man moves, walks, but does not talk. The work is flawed because of some inequity that must exist in Rava, the creator. The creation is undone, sent back to dust. In commentary on the text, Rashi notes that this sort of magical enterprise (in a way, the basic science of the time) was achievable by the manipulation of the letters of the name of God, the building blocks, as it were, of the Creator, as known by humans. The commerce is language: the word and the letters. In later Golem tales (and the legend persists),[54] the Golem has the Hebrew letters of the word "truth," *emet*, carved on his forehead. By removing the aleph, the first letter in one of God's names, the Hebrew word "death," *met*, is formed instead; and the Golem vanishes. Further legends link the Golem not only with the chimera of truth but play with the Golem: all body, no spirit. The Golem in later tales is a revenging and powerful force: unlike the caricatured and vilified Jewish body, small, stooped, and awkward, the Golem of Prague legend is tall, muscular, and powerful—wreaking havoc on the Gentile enemy. The Golem of Prague emerges to protect the Jews from the wrath of the Gentiles on the eve of Passover 1580, when the blood libel charges historically increased and led to pogroms.

Yet as appealing as this image is to a persecuted people, we are warned of the essential error in the pursuit of this particular type of creationist research: the manipulation of the whole by pieces of the whole leads in these texts not to truth, but to the excesses of spiritless power, unguided by faith, ultimately dangerous. The texts are cautionary but apparently not absolutely prohibitive; otherwise, the persistence of the story would not be evident.

The themes, I would argue, offer some measure of lesson in the search for a theological reflection on the Human Genome Project itself.[55] Here are texts that speak of the limits of the human creative power, of the caution that ought to attend to its use, of the care that must be exercised in the face of the possible act, and of the hiddenness of inequity in the human soul that leads to all manner of unexpected and unforeseen losses in the final result.

The Golem texts do not prohibit the HGP any more than does the contextual history of the tragically and inevitably bad science of the nineteenth century. But taken together, they do testify to the necessity for extraordinary cause for concern, and the textual basis for a framing story of worry. Anyone who takes seriously the potential to cure disease is affected by the promises of the scientist. It is folly to insist on diversity of the human species over the bedside of a child dying of Tay-Sachs disease, for example. But at this time, the only "solution" to the discovery of the "abnormal" is the elimination of the person who would carry that trait, and thousands of other traits, into our world.[56] It is a terrible moment to be in this science. A current story illustrates this complexity. The gene for cystic fibrosis was one of the first to be mapped and located by the project, to the enormous fanfare of the scientific community and the great relief of the families of the children who suffer from a disorder that, in its worst form, leads to an early childhood death from the tenaciously thick mucoid secretion that blocks and eventually pathologically scars the lungs and pancreas. In Great Britain and Canada, large population screening was set up to administer the prenatal diagnostic testing to prospective parents, and on finding that their child carried the genes for cystic fibrosis, abortion was offered. But the screening revealed that cystic fibrosis in a variety of nonlethal forms was far more prevalent than had been suspected. And in November 1993, the *New York Times*[57] reported on the confusion about the meaning of these findings under the headline: "Cystic Fibrosis Surprise: Genetic Screening Falters." It turns out that the gene's true placement is of limited help. Some forms of cystic fibrosis manifest only as adult male infertility, others as lethal. From a positive cystic fibrosis finding, it is impossible to tell how the condition will manifest. And further, there is discomfort at the implication for the complete elimination of the cystic fibrosis gene, or gene pattern. There is speculation, for example, that its widespread occurrence might have had some epidemiological or adaptive function, as in sickle-cell carrier status and malaria protection—that carriers were less susceptible to the infectious diseases of the northern European countries where tuberculosis or other respiratory conditions are most common.[58]

Yet the very next day in the *New York Times*, the headline read: "AMA Is Heralding a Revolution in Genetics."[59] With no mention of the previous day's "surprise," the article summarizes recent work published in the *Journal of the American Medical Association* (*JAMA*) that "the genetics revolution is well under way . . . an explosion, a culmination, a profound statement of the importance of this subject for the human condition."

It is not only deficit disorders that this new science will uncover, but various mental states, including depression, the modern parallel to the diagnosis of hysteria in an earlier time, a disease diagnosed and treated nearly entirely in women:

> Advances in genetics have made it feasible to find genes that predispose a carrier to common diseases like heart disease and mental disorders, manic depression and schizophrenia . . . not since the [discovery of germs theory] has the stage been set for as radical rethinking of the causes for disease and disability as is now occurring with the elucidation of the human genome.

Meanwhile, on the editorial page the same day in the *New York Times*, biologist Robert Pollack writes about the application of such theoretical research, and location and identification of genes are already taking place. Mice and farm animals are regularly altered and "manufactured" by the use of transgenes, the insertion of a specifically chosen, desirable gene into the primary cell of the organism. It is theoretically possible to do the same for humans, he states, but there are problems: the first experiments, like all scientific experiments, may be flawed. In this case, however, the first experiments will be living babies; persons that we will in some sense assemble with the genetic letters of the creative instructions. Pollack calls for congressional hearings on the topic.

Conclusion

I contend that we will need far more. The discourse of bioethics and Jewish ethics needs to address this task with utmost urgency and with a sober focus on the potential for peril as well as glory. In this country, visible difference is used as the signifier of not only difference, but also exclusion. Such exclusion has demonstrable health effects: it is not simply a matter of theoretical injustice. In the early half of the century, immigrants, notably Jews, were marked in this way, and throughout the history of this country, African Americans were most clearly marginalized by the sign on the body. But the thin delineation of the human also exists as a social mechanism to exclude elderly persons, children, and people who are disabled from the productive and hence efficacious community. The sign of exclusion is increasingly a sign not of our increased responsivity to help, but of an increased tendency to leave outside the discourse, to step over in the street, to "exclude from the benefit package" in the newest language of health care reform. The power to describe and define the normal human is itself a normative act: it derives its essential meaning from power relationships and truth claims that then construct the possible world. Hence, while it is inevitable that the Human Genome Project will proceed, and it is likely that the genetic model will replace the germ theory as the model for causation, and while it is undeniable that great good most likely will occur, it must proceed within a community that is fully aware of its history and of some essential human obligation for one to another.

"The Genome Project is a glorious good. . . . I have spent my career trying to get a chemical explanation for life, an explanation for why we are human beings and not monkeys."[60] These are the words of James Watson, first director of the HGP and its most enthusiastic defender. He contrasts in a paper devoted to this defense the real search for the "true" origins of life and the meaning of human existence with the "simplistic" answers of Revelation, the "recourse" to God. But it is only by bringing the full and troubling context of a faith commitment to this work that can fully illuminate it. Using the text in this way is not simply a quaint folkloric curiosity. It is only by insisting that the public discourse about the common good is barren without the perspective of the faith community that such discourse can proceed.

The power of the perspective of science is enormous; so much so that in our time, the clinical clearly dominates and defines the discourse itself, and the search for efficacy and consequence can seem our only rational task. But fidelity, justice, obligation, and the persistence of history must call us to another kind of attention. To discuss the HGP within a community of moral meaning and purpose will mean hearing about the nature of suffering and the reality of limits. What makes us human is precisely this discursive community: fallible, difficult, yearning, and capable of both great vision and great evil. From the Jewish perspective, this will mean the source texts will be made not only of the *halakhic* codes and *responsum,* but also the cautionary *midrashic* admonitions, and perhaps most centrally, from the blood and bones and flesh of our history. The map of the normal human self, pieced together, is at last not the key to meaning; it is the unvoiced collection of the letters without the voice of the human. At issue for the public discourse is the same concern that is expressed in the Talmud. If one is to create a world, the difficult part is not the technical; it is the "inequities between you and your God." It is the issue of righteousness and justice of the ones who would be world creators that is at stake. Unless we regard such work with care, extreme care, then the truth promise on the forehead of the Human Gene Project will shift before our eyes (without the aleph, *emet* will be gone, only *met* remains), and transform the work into dust in our own hands.

Notes

1. The goal is the location and charting of all the genes of a human karyotype on the chromosomal structure itself. The knowledge obtained by the completion of the project is promised by the scientists funded for the research to be revolutionary.
2. Daniel J. Kevles, "Out of Eugenics: The Historical Politics of the Human Genome," in *The Code of Codes,* ed. Daniel Kevles and Leroy Hood (Cambridge: Harvard University Press, 1992), 3.
3. Kevles and Hood, *The Code of Codes,* viii.
4. Ibid., vii.
5. This chapter represents one personal perspective on the Jewish historical and textual tradition. There can be many others, diversity and lively dialogue being characteristics of Jewish philosophy. There is no claim that this is *the* Jewish view. My thanks to Solomon Katz for this point.
6. The "Shoah" is the modern Jewish term for the Holocaust. It will be used throughout this chapter.
7. Emmanuel Jakobowitz, both in his authoritative book *Jewish Medical Ethics*

(New York: Bloch Publishing House, 1959) and in speeches in 1990–93, has referred to the medical moral imperative to heal as most profoundly addressed by the HGP.

8. Many texts focus on the Jewish male. Some included the Jewish woman in the term "Jew," but most did not, paralleling the assumption that the Christian body was the male body as well. In this discourse, as Daniel Boyarin has noted, the Jewish male was "feminized," or given female attributes as a further denigration of his condition.

9. Sander Gilman, "The Jewish Foot," in *The Jew's Body* (New York: Routledge, 1991), 39. Gilman's work, not directed at all to the HGP, but in the field of Jewish studies and culture, is the single most critical text in understanding the history of the mapping of difference for the Jew. His work is invaluable to the problem of science and the genome. My thanks to him for providing the historical text to support what had been only a vague sense of uneasiness on my part.

10. Ibid., chaps. 2 and 7.

11. Barbara Tuchman, *A Distant Mirror* (New York: Knopf, 1978), 111.

12. Gilman, *The Jew's Body*, 39.

13. Ibid., plate ten, 50.

14. To my knowledge, there is nothing to suggest that religious or intellectual Jewish leaders challenged this view either.

15. The Zionist movement actually internalized some forms of this in the ideological commitment to the creation of a "new Jew"; the stress on gymnastics and hard physical labor were to transform the body of the Jew.

16. The work of Knox is quoted by Gilman, *The Jew's Body*, 174.

17. Ibid., 76.

18. Ibid.

19. In the American Holocaust Museum, there is elaborate testimony to this, with exhibits showing the cards of normal human hair, nose charts, and calipers for measures of the normal. The science of eugenics reached intricate heights in the prewar period. Later exhibits continue to document the Nazi fascination with this issue, marked by the death camp experiments with "normal human limits."

20. "What makes contemporary political democracies unlikely to embrace eugenics is that they contain powerful anti-eugenic constituencies. Awareness of the barbarities and cruelties of state-sponsored eugenics in the past has tended to set most geneticists and the public at large against such programs." Daniel J. Kevles and Leroy Hood, "Reflections," in *The Code of Codes*, 318.

21. Troy Duster, *Backdoor to Eugenics* (New York: Routledge, 1990), 9–13. Duster is critical. The air force attempted to exclude African Americans with sickle-cell carrier status from piloting jet aircraft (followed closely by the commercial airlines), after a discovery that sickle-cell disease, in some isolated

cases, could cause loss of consciousness in certain circumstances of low oxygenation. It was a haunting parallel to the bad soldier–flatfoot analogy of the Prussian army of the 1800s. It was rejected as social policy only when it was shown that not only did the carrier status have little to do with the symptoms described, but that a dramatic proportion of NFL players, who routinely flew and played football at high altitudes, suffered no ill effects.

22. Ibid., 11. That this link is present and commonly made does not escape the attention of modern theater. In a recent play on Broadway entitled *Twilight of the Golds*, characters argue about this problem. The characters are Jews, in fact broadly caricatured Jews with oddly anachronistic mannerisms of the 1950s, faced with a dilemma of the 1990s: whether to abort a fetus who is a carrier of the "gay gene." Nobody in the play turns to spirituality or faith as a guide; Wagner, the legendary anti-Semite, dominates the music and metaphor of the play. The characters, however, do discuss their Holocaust history and how it troubles the geneticist husband to work on a project that charts "the normal." In this play the Jewish gay male is "feminized" and diseased, and the fetus is the stand-in for the Jew of the nineteenth century.

23. Such nativist sentiment and these same laws blocked the legal immigration of thousands of European Jews threatened ten years later by Hitler.

24. Kevles and Hood, *The Code of Codes*, chap. 1. See also *Gene Mapping: Using Law and Ethics as Guides*, ed. George Annas and Sherman Elias (New York: Oxford University Press, 1992). Many of the authors discuss the problem mentioned in this chapter, as well as many others, yet most seem assured that direct eugenics links by the government are unlikely.

25. The Sunday *San Francisco Chronicle and Examiner*, in the aptly named "Image" section. This example appeared weekly during the period between July and October 1993.

26. For the purposes of this published text, this name has been altered from a name with the same derivation. This name and others like it derive from the Hebrew *Cohen*, meaning "priest" and suggesting affiliation with the priestly class in Jewish tradition. To this day in traditionally observant religious liturgy, the Cohenim have special ritual obligations and are prohibited in some contexts from other obligations.

27. Many other current examples abound. In articles appearing in the *Washington Post* and the *New York Times*, which examined the ideas and influence of Michael Lerner, a Jewish intellectual and editor of a leading Jewish journal called *Tikkun*, the writers actually ridiculed Lerner's "big feet," "shuffling gait," "overly eager manner," "shifting gaze," and short stature relative to the tall, cool Gentile Clintons.

28. As is abundantly true in veterinary medicine already.

29. Daniel Boyarin, *Carnal Israel* (Berkeley: University of California Press, 1993), 1. Also noted in his forthcoming work on Freud, and in private conversation. My gratitude to Professor Boyarin.

30. Jared Diamond, *Natural History* 102, no. 11 (November 1993): 12–20.

31. Ibid., 16.

32. Ibid.

33. Ibid., 18.

34. Ibid., 16.

35. Al Jonson, as quoted in the *Los Angeles Times*, November 1, 1993, A1, A12.

36. This is a tricky point in scholarship. What I am suggesting is that this perspective is not the special responsibility of Jews. The history of the malevolent encounter is, in part, the shared history of Western medicine. When Jewish scholars read the material, they need to place the account of what occurred to our people as central to the discourse of the *halakhah*.

37. *Compendium on Medical Ethics*, ed. David Feldman and Fred Rosner, 6th ed. (New York: Federation of Jewish Philanthropies of New York, 1984), 12.

38. Talmud Bavli, Bava Kamma 85a.

39. Talmud, Sanhedrin 73a.

40. Jakobowitz, *Jewish Medical Ethics*, 22.

41. Elliott Dorff, *Choose Life: A Jewish Perspective on Medical Ethics*, IV:I (University of Judaism: University Papers, February 1985). Here he is quoting Midrash Temurrah as cited in Otzar Midrashim.

42. Ibid., 16.

43. Talmud Bavli, Tractate Niddah.

44. David Bleich, *Judaism and Healing: Halakhic Perspectives* (New York: Ktav, 1981), 106.

45. David Bleich in private conversation. My gratitude to Professor Bleich for his assistance in thinking about these issues. While I tend to worry about the history, I am assured by his expertise that the *halakhic* texts do not exist to ground such worry.

46. Felice Maranz, "In Whose Image?" *Jerusalem Report*, September 23, 1993.

47. Ibid., 14. See also Gina Kolata, "Nightmare or Dream of a New Era in Genetics?" *New York Times*, December 7, 1993, A1.

48. Maranz, "In Whose Image?" 13. The *Jerusalem Report* is the largest English-language journal in Israel and is widely distributed in the United States. In this cover story, a variety of *halakhic* authorities are quoted as cautiously approving the policies. As I noted above, the texts support such an approach. Rabbi Moshe Tendler is mentioned as raising serious oppositional questions. Dr. Tendler's most recent work is as yet unpublished, but at the time of this writing, I am in correspondence with him about the textual sources for this opposition.

49. Kolata, "Nightmare or Dream," B10.

50. *Daily Mail of London*, "Science" section, July 1993.

51. Immanual Jakobowitz is the one of the most respected *halakhic* authorities in

the field of Jewish medical ethics in the modern era. His *Jewish Medical Ethics* was the first of its type in this field. His remarks must be understood in this context, which is why their publication was so widely reported in the international press. This statement was later clarified by the current chief rabbi of the United Kingdom, who issued his own statement and noted that homosexuality was rather too complex to be removed or produced by any scientific procedure.

52. I was well into the work of this project when I came across a reference to the very excellent article about not only the dangers of the early work in genetics, but in the creation of artificial intelligence and machines that mimic human capabilities. Byron Sherwin, "The Moral Implications of the Golem Legend," in *In Partnership with God: Contemporary Jewish Law and Ethics* (Syracuse: Syracuse University Press, 1990), 181–208. Sherwin's work is in many ways prophetic. He is writing before the HGP is under way, but worries about genetic research in general as an enterprise akin to the creation of the Golem. This chapter is a good review of the Golem legends, a summary that for the purposes of conserving space I did not include here.

53. Talmud, Sanhedrin, 65b, as quoted in Moshe Idel, *Golem* (New York: State University of New York Press, 1990), 27.

54. It is a subject of a novel by Marge Piercy, *He, She, and It*, published in 1992. The tales recur in the eighteenth century, and in the texts of the *responsa* literature. See Zevi Askenazi, *She'elot u-Teshuvot*, no. 93. In one such text, the question is raised about whether the Golem can be included in a prayer quorum, a *minyon*. At stake is the issue of murder. If the Golem is a man, then is it not killing to "return him to dust"? (One thinks here of the legal cases involving the destruction of embryos.) The text resolves this in an odd way, not by claiming the humanity or countable status of the Golem, but by decrying the waste of a creature with "a purpose." Sherwin also comments on this text and notes that Askenazi's son, Jacob Emden, argues with this distinction. Emden is an important commentator on other issues in medical ethics.

55. One could reflect on other texts as well. The story of the Tower of Babel is a description at least in part of an earnest human endeavor similar in tone to the international project of genome mapping. The destruction of the tower—and the destruction, far more significant, of language held in common—is troubling to the rabbinic commentators. Why such a response? The text does not tell us; the rabbinic Midrash does: in building the tower, the brick and the technology were more highly valued than the human persons doing the work. The brick took a long time to make and to haul to the top of the tower, and when it broke, the people mourned; but when a man fell, they paved over him and went on.

56. A personal note: During the last week of writing this chapter, I typed these words as I awaited the results of the prenatal testing done in the third month of my pregnancy. I am a healthy, forty-three-year-old woman; this is my fifth

child; there is no reason for alarm. But I waited in terror for the phone call that would tell me whether my child was "normal" or "abnormal"—of the human community, by implication, or outside, disabled.

57. *New York Times*, "Science," November 16, 1993, B6. Of interest is that this day's news was marked by three stories of medical failure: the differences in care for African American women after hysterectomy surgery, and the failure of the FDA to regulate an experimental drug resulting in the deaths of five people. It is instructive how much medical error and mismanagement, not to speak of overt institutional racism, we can accommodate and still speak of the neutrality and objectivity of basic research.

58. Tay-Sachs is also speculated to have an adaptive role for Jews in preventing tuberculosis. Both comments from Benjamin Wilhelm in private conversation.

59. *New York Times*, "Health," November 17, 1993, B7. Both stories were written by Gina Kolata. "Beyond Cloning," Robert Pollack, A19. On November 17 the *Times* was considering the genome from all angles: even the food editor, Jean Brody, had an article about genetically engineered milk and the attendant fears about its safety, on B7.

60. Kevles and Hood, *The Code of Codes*, 164.

PART THREE

SOCIAL CHALLENGES

[9]

Risk Classification, Genetic Testing, and Health Care: A Conflict between Libertarian and Egalitarian Values?

David A. Peters

GENETIC TESTING IS A powerful new tool for predicting the likelihood of future health problems for individuals. Three important types of genetic information are derivable from genetic testing. First, genetic testing can reveal whether individuals carry recessive genes for various heritable disorders transmissible to progeny (e.g., Tay-Sachs disease, sickle-cell anemia, etc.). Second, genetic screening can provide presymptomatic diagnosis of the presence of a number of strongly determinative genes (typically single defective genes) for certain disorders, among them Huntington's chorea and adult onset polycystic kidney disease. Individuals carrying genes for these disorders *will* develop the disorders if they live long enough. Third, genetic testing can identify the loci of certain so-called contingency genes (usually a plurality of interacting genes) in individuals that predispose them to disorders that will become manifest *if* other genetic and/or environmental conditions obtain. Contingency genes for abnormal cholesterol levels, arthritic and autoimmune phenomena, and manic-depressive illness have already been discovered. More than three hundred markers for contingency genes have been identified,[1] and the pace of discovery can be expected to increase in the future.

What might this powerful new tool of genetic testing mean for health care in the United States? It might mean that some persons will test right out of access to health care. The task in this chapter is to show the likely impact genetic testing will have on our insurance-based method of providing health care. In showing this, I would like to show also that we in the United States approach such matters from two different, yet coherent underlying schools of ethical thought, namely, the libertarian and the egalitarian. Both libertarianism and egalitarianism affirm justice, but justice looks different in each case. The advent of genetic testing will smoke out these differences.

Might We Lose Our Health Insurance?

Continuing research will doubtless expand the power of genetic testing to predict future illness. Consequently, such testing will become more widespread in clinical settings for the purposes of prevention, early diagnosis, and early therapy. The results of these tests will be recorded in patients' medical records. Commercial health insurers will obviously be interested in—and may demand—access to genetic test information in reviewing applications, especially for individual health insurance policies.[2] It is plausible to believe that insurers will view positive genetic test results for various strongly determinative genes or contingency genes like those described above as relevant factors in assessing an individual's composite risk profile. Armed with such data, insurers may categorize more individuals as high risk than now receive this designation. Following current practice, insurers will then likely either deny coverage to such individuals[3] or accept them at elevated rates—rates that may be so high that many high-risk applicants may be unable to afford them, leading them to "choose" to go without coverage. In this manner, genetic testing could indirectly contribute to the total number of people lacking health insurance in the United States. And since health insurance is the typically necessary condition for receiving medical care in this country, genetic testing could thus indirectly contribute[4] to increasing the number of individuals lacking access to adequate health care in the event of need. This is the often voiced worry concerning the dark side of the development and proliferation of genetic testing.

Can We Blame the Insurance Industry?

The hypothetical causal sequence just described may at first sight make the underwriting practices of commercial insurers (i.e., the practices of risk classification and prorating of premiums to risk status) appear to be morally suspect so far as they contribute to the major problem of inequitable access to health care among people in the United States. The insurance industry would thus seem to bear a burden of proof to morally justify these procedures. Insurers are able to respond to this challenge, however, with a variety of arguments. I shall here focus on one of their standard lines of defense—a defense that appeals overtly to a notion of "equity"[5] (justice), which the industry claims is at the heart of its practices of risk rating and the calibrating of premiums to risk status.

Formally (i.e., abstractly) stated, the principle of equity (or justice) is this: equals should be treated equally. A longer and richer formulation of the principle runs this way: similar things should be treated similarly and dissimilar things treated dissimilarly in proportion to their dissimilarities. Applied to the marketing of health insurance, this principle requires equal pricing of insurance among those of equal risk (horizontal equity) and differential pricing of insurance among those of unequal risk (vertical equity), according to commercial insurers. Following Daniels,[6] let us call this substantive notion of equity (justice) embraced by insurers the principle of "actuarial justice." For example, reliable statistics support the claim that someone with coronary heart disease or a recent history of cancer is at greater risk of future illness and its associated medical expenses than someone who has not sustained these medical problems. Insurers argue that persons with more serious risks to health are getting protection against a greater level of likely medical expenses than persons with less serious risks. Therefore, it is only "fair" that "greater risk" people pay more than "lesser risk" people. Moreover, say insurers, if individuals and groups are not segregated into different insurance pools according to risk status ("standard" or "preferred" risks versus "substandard" risks)—as would be the case under so-called community rating where everyone pays the same premium irrespective of risk status—then low-risk people would end up subsidizing high-risk people. Insurers assert that the problem with such cross-subsidization is twofold. First, it is clearly not in the objective best interests of low-risk insureds. After all, in a competitive insurance market they *could* be paying a lower premium commensurate with their lower risk status if they were separated from the high-risk individuals and consolidated into a discrete pool for rate-setting purposes. Second, low-risk individuals would not likely agree to community rating if they understood what was financially at stake for them. (There is ample evidence to support this latter claim).[7] Thus, the involuntary merging of high- and low-risk people into common pools under, say, government-mandated community rating would violate the formal principle of equity substantively interpreted as the principle of actuarial justice. Under community rating, an equal premium would be charged to individuals who are in fact unequal in risk status.

The Libertarian Values of the Insurance Industry

I argue that the commercial insurers' defense of risk classification and the calibrating of premiums to risk status rests on a set of claims that can be identified as essentially libertarian in substance.[8] Two libertarian theses (arguments) undergirding the insurance industry's justification of these practices are particularly relevant to the present explication and subsequent critique of their defense.

1. Freedom to act in accord with one's choices—so long as one does not infringe on the like freedom of others—is *the* supreme value, a person's most fundamental "natural right," a right that "trumps" all other moral considerations. This general right implies a cluster of more specific so-called negative rights (e.g., the right *not* to be killed or assaulted, the right *not* to be coerced, the right *not* to have one's property taken or destroyed, etc.). These rights place upon others correlative *duties* that are also negative in nature so far as they require various forms of noninterference by others (e.g., the duty *not* to kill or assault anyone).

2. Freedom to do with one's property (income) as one sees fit is a necessary and supremely important condition for living life as one pleases. Therefore, everyone has a (derivative) "natural right" to dispose of his or her fairly obtained income as he or she desires. It follows that civil government is never morally justified in forcing any citizen, against the person's will, to part with any portion of his or her legitimately gained income (e.g., through taxation) to support projects or programs the person does not approve of.

In connection with thesis 2, libertarian Robert Nozick would likely offer the following "buttressing" argument explaining the thesis's relevance to the issue of insurers' use of genetic test results for the purposes of risk classification and rate setting.

Nozick would concede that the natural "distribution" of "good" and "bad" genes related to future good or ill health is arbitrary. Some people are "winners" and others "losers" in nature's "genetic lottery." A person who is relatively free of genes linked to the onset of future medical problems, or who is free of genes that otherwise would have caused congenital disabilities, admittedly does not "deserve" his or her "good" genetic makeup. Nonetheless, Nozick would say, the person "has" this "healthy" genetic constitution, and "*not illegitimately*,"[9] that is, the manner in which the person "acquired" his or her particular genetic makeup violated no one else's prior property interests in these same genes. A person's genetic makeup is his or her "natural property." Therefore, if commercial insurers freely offer to such a person a lower premium rate because he or she has tested negative on certain genetic tests predictive of future health problems, then that person is entitled to this voluntarily conferred economic benefit. And, Nozick would maintain, this "natural" entitlement should be protected by the coercive power of the state.

If this conclusion is denied, Nozick would point out, then one will have to say that any person who reaps economic benefits from others for traits or for the use of traits (e.g., talents, abilities) that are *genetically based* (e.g., the physical beauty of a professional model, the athletic prowess of a professional basketball player, the engaging voice of a radio or TV announcer, the mental acumen of a great mathematician, etc.) is not *entitled* to these benefits (and to state protection of them) because the traits that are used by the benefited party are not *themselves* deserved. This position is counterintuitive and would force us to conclude that no one is entitled to any economic benefit others freely bestow on those who have and use valued traits that are genetically based. But this is an absurd position. So if commercial insurers, without coercion, give economic benefits (in the form of lower premiums) to people with "health maintaining" genes, but do not choose to give this benefit to those with "health jeopardizing" genes, it is thoroughly consistent with our normal intuitions to say that the former are *entitled* to the economic benefits they receive from insurers for their low-risk genetic constitution. It follows that the state should give this "natural" entitlement the protection of law. Nozick says, "Whether or not people's natural assets are arbitrary from a moral point of view, they are entitled to them, and to what flows from them."[10]

Critique of the "Genes as Natural Property" Argument

According to libertarian thesis 2, property rights are important because they are essential means for practically exercising a person's fundamental right to make a life for oneself through uncoerced personal choice. But even if this point is conceded, it leaves entirely open at least two important questions:

1. What specific sorts of things should the state recognize as legitimate objects of property interests?
2. For the things acknowledged as legitimate items of property, what should be the *specific scope* of the liberties, powers, and claims granted by the state to the "owner" of this recognized property? (The decision made on this issue will affect the scope of the correlative duties [restraints on freedom] placed upon others vis-à-vis the liberties, powers, and claims granted to the recognized "owner" of the property in question.)

The "should" in both questions is a *moral* "should." As examples of the first question, consider the following: Should a woman have property rights in a fetus she has aborted—such that only with her express consent can brain cells be removed from the abortus for therapeutic transfer to the brains of individuals suffering from Parkinson's disease? Should a person be given property rights to his or her own kidneys—permitting the person to sell a kidney to someone dying of end-stage renal disease? Specific examples of the second question might be: Even if a person is ceded property rights to something (say, a parcel of land on which his home is located), should the scope of the person's exclusive right of use of this property extend so far as to permit him to raise pigs in his backyard, or to permit him to refuse use of this property by others for *any* reason, for example, as part of a projected city bypass route?

These questions show that whether the state *should* recognize property interests in certain things at all, and what *specific* liberties, powers, and claims *should* be given state protection in connection with some item allowed to have property status, can be determined only by reference to other *moral* considerations beyond the protection of personal liberty. These other moral considerations might plausibly include the principle of utility (promote the greatest good of the greatest number), the duty to protect the vulnerable from harm, other substantive notions of justice besides "actuarial justice," and so on. The point is that the idea of "natural" property interests in things, that is, property interests that "predate" the establishment of civil government, is a myth. To say that civil government "ought" to recognize property interests of some specific sort in some type of thing is to imply that this "ought" claim can be justified ultimately by appeal to general moral principles that are substantively neutral concerning "property matters."

Thus, whether a person's "good" genetic makeup (i.e., relative lack of health-threatening genes) *should* be recognized as his or her "property," *entitling* the person to state protection of the economic benefits bestowed by insurers (in the form of lower premiums), *depends* on whether giving legal recognition to such property interests is justified, all things considered, by appeal to *moral* considerations *beyond* the mere protection of personal liberty, libertarianism's exclusive concern. Given this contingency, it is an open question whether government-mandated community rating for health insurance policies—which would deny insurers the liberty of offering lower premiums to those at lower genetic risk of disease, and would simultaneously deprive genetically low-risk individuals of the economic benefits insurers currently offer such people—*is* morally wrong, despite the fact that such a government policy would obviously violate the principle of actuarial justice (which is, arguably, not a *moral* principle at all).

Egalitarianism, Genetically Based Health Risks, and "Just" Health Insurance Policy

In a manner parallel to that of Karen Lebacqz, who in an earlier chapter distinguished between a Lockean and a Lebacqzian vision of fair shares, I would like at this point to explore a nonlibertarian matrix of thought, namely, egalitarianism. I would like to weigh the alternatives and then move toward some conclusions.

How does egalitarianism assess the moral propriety of calibrating premiums to health risks grounded in an individual's genetic inheritance? The social policy debate on this issue will, I suggest, reflect the presence and tension in each of us of both libertarian and egalitarian commitments.[11]

The foundational premise of egalitarianism is that every human has supreme dignity and moral worth insofar as he or she possesses the general capacity for critical reflection, which includes the specific capacity for moral autonomy. The general capacity for critical reflection refers to the ability normal humans have to subject their own beliefs, arguments, and actions to critical scrutiny, to adopt a "spectator's" view with respect to these items, and to ask questions such as: How good are the reasons that support the factual beliefs in my belief set? Are the arguments I am prepared to adduce in defense of certain claims sound and logically valid? The specific capacity for moral autonomy refers to the capacity to ask questions such as: Is an action I performed (or am contemplating doing) morally right—all things considered? What moral rules and principles ought to guide human action?

Our capacity for self-examination in the nature of the case gives pride of place to reason. While emotions and appetites admittedly incline us to accept certain beliefs and arguments and to act in certain ways, our reason-centered capacity for critical reflection demands that these beliefs, arguments, and actions ultimately be justified with reasons that are logically consistent with all the other beliefs (factual and valuational) comprising our total belief set at any given time. This deference to reason over the emotions and appetites is in a sense self-certifying. Reason will demand good reasons for abandoning reason as our primary guide to action, or for relegating reason to an inferior status vis-à-vis the emotions and appetites. This obviously is an unachievable task, however, since it involves conceding the primacy of reason (in the form of conceding the legitimacy of the demand for good reasons) for the presumed purpose of demonstrating reason's supposed inferiority to these nonrational impulses regarding belief and action.

Our capacity for critical reflection has a built-in thrust or aim, namely, to transcend or get beyond all limitations placed on reason's critical operation by ignorance, superstition, prejudice and, perhaps most important, the limitation posed by the desire to gain and maintain personal advantage over others. With respect to this latter desire, reason will show its independence and ever critical orientation by querying: *Why* should *my* interests take precedence over the interests of others? The very question betrays the point of view that the capacity for critical reflection aims at, which is its built-in ideal, namely, the impartial point of view. From the impartial point of view, a just law, policy, institution, or practice is one that a person can defend to himself or herself and to others without appealing to personal advantage. A just person is one who desires to act in ways that can be defended to himself or herself and others without appealing to personal advantage.

These foundational claims of egalitarianism imply certain others that, when supplemented by a number of defensible factual claims, together warrant the conclusion that access to adequate health care in the event of need is a social good that (morally) should be available to *every citizen irrespective of his or her ability to pay*. If in the United States health insurance continues to be a necessary condition for receiving adequate medical care, then egalitarians will contend that all individuals with the capacity for critical reflection and moral autonomy have equal claim to health insurance sufficient to provide them with an adequate level of health care—if such care can technically be made available to all in society. If adequate medical care can be technically provided all citizens, then from the egalitarian perspective, the current circumstance of thirty-five million people without health insurance in this country is not merely an unfortunate state of affairs for those left uninsured (as libertarians would describe it). It is matter of *injustice*. It constitutes a *moral* problem, which the state is duty bound to rectify.

Details of this general egalitarian argument need to be filled in, however. According to egalitarians, all individuals are moral equals just so far as all possess the capacity for rational self-examination. This general capacity includes the capacity for critically reflecting on the morality of one's acts, and at an even deeper level it involves the capacity for critically assessing the tenability of individual alternative moral rules and principles, and entire sets of such rules and principles. Equal possession of the capacity for critical reflection and for moral autonomy is for egalitarians the basis for asserting that each individual should be legally free (i.e., that each individual should have the legal right) to live his or her life as he or she sees fit, so long as the person does not interfere with the like freedom of others. Practically speaking, for egalitarians the aforementioned right refers to the state-protected freedom every individual should have to make a life for himself or herself in terms of the set of life options (roles, vocations) that comprise the "normal opportunity range"[12] for any individual in society with his or her talents, skills, and abilities.

Modern egalitarians such as John Rawls[13] and Norman Daniels[14] acknowledge, however, the obvious fact that humans are not just disembodied streams of thinking and choosing. We are physical bodies subject to illness, dysfunction, and disability. These physical pathologies are morally important because sickness and disability can compromise, first, an individual's efficient exercise of his or her most important capacity for critical reflection and moral autonomy (physical pain can distort clear thinking). Second, illness and disability can place constraints on a person's practical ability to make a life for himself or herself from the array of life options in society's normal opportunity range (i.e., illness and disability can jeopardize "equal opportunity").

For these reasons egalitarians assert that access to adequate medical care in the event of need is as vital for one morally autonomous person as another. And since all morally autonomous individuals have equal moral value by virtue of their possessing the capacity for critical reflection and moral autonomy, it follows that if preventive or therapeutic medical care is (or can be readily made) technically available to all morally autonomous individuals, then the nonequal distribution of such care among needy equals according to some arbitrary criterion like ability to pay is unacceptable. That is why egalitarianism requires *universal* access to adequate medical care, *irrespective of ability to pay*. Egalitarianism thrusts on libertarian-oriented commercial insurers a burden of proof to morally justify risk classification and the prorating of health insurance premiums to risk status—*if* these risks are arguably not brought about by an individual's voluntary action. Mere possession of strongly determinative genes or contingency genes linked to the onset of future illness would be model cases of such involuntarily assumed risks.

Egalitarians would assert that the libertarian-based claim of commercial insurers—the claim that individuals are entitled (read *morally entitled*) to the economic benefits the industry confers on those with low genetically based risk of future illness—would *not* be unanimously agreed to by all individuals who assess this practice *simply* from the perspective of someone endowed with the capacity for critical reflection and moral autonomy, bracketing from consideration all other descriptors (e.g., skin color, sex, genetic makeup, etc.). For egalitarians, these latter traits (and the "favoritisms" [prejudices] often allied with them) are irrelevant distinctions between individuals for the purpose of determining a "just" distribution of certain critical social goods, such as adequate housing, nutrition, medical care, and so forth. These traits are irrelevant for the distributional purposes cited because they would be so regarded from the *impartial point of view*—the point of view aimed at by the capacity for critical reflection itself.

Since the real-world counterpart of the economic *benefits* conferred by insurers on congenitally low-risk people (i.e., lower premiums) is an increased economic *burden* (i.e., higher premiums) charged to congenitally high-risk people (these burdens can create medical poverty[15] and/or lack of access to adequate medical care in case of need), the impartial point of view cannot certify the latter "surcharge" as "fair." As we have seen, however, from the libertarian-oriented commercial insurers' position, charging higher premiums to higher risk people *is* fair. Insurers insist that the practice is a form of treating equals equally and unequals unequally in proportion to their inequalities.

The crucial distinction between libertarians and egalitarians is that the former do, and the latter do not, regard differences in genetic makeup among individuals (who are otherwise equally endowed with the capacity for critical reflection and moral autonomy) as a morally relevant basis for distributing the economic burdens and benefits connected with health insurance premiums. Thus, egalitarianism strongly inclines toward *community rating*—at least for all health risks that are not voluntarily assumed or controllable by the individual without "undue" cost, effort, or inconvenience. (The issue is more complex, however, when attention is shifted to health risks that do *not* seem to be mere dictates of fate, but are more directly related to the voluntary behavior of individuals. These are often called lifestyle-generated risks.) It seems fair to say, then, that libertarians view the function of insurance from a narrowly individualistic, self-interested perspective as a means of personal risk management. Egalitarians, on the other hand, tend to view the function of health insurance from a social perspective as an instrument for protecting the efficient exercise of critical reflection and moral autonomy and for guaranteeing equal opportunity as earlier defined.

It bears noting in closing, however, that while egalitarianism requires as a matter of justice that there be universal access to medical care in the event of need—irrespective of ability to pay—it does not uniquely require any particular social arrangement for achieving this goal. Egalitarianism does not dictate the establishment of a national health insurance system on the Canadian model. Nor does it necessarily call for the development of a national health service as in Great Britain. Indeed, egalitarianism does not require that health insurance be completely removed from market competition among private insurers. Such competition offers certain advantages less likely to be available under entirely government-run national health insurance or national health service programs: different benefit packages that cater to a diversity of preferences,[16] opportunities for experimentation and innovation, and incentives for more efficient performance.[17] Managed competition between various health plans all striving to attract the business of large regional insurance-purchasing co-ops— the centerpiece of the Clinton administration's failed proposal for health insurance reform in the United States—is not inherently inconsistent with the requirements of justice set forth by the egalitarian position.

Conclusions

The continuing development of genetic testing capable of predicting future health risks for individuals carries the promise of allowing earlier diagnosis of disease and earlier preventive and therapeutic interventions. However, the technology has a potential dark side. Genetic test results may be used by commercial insurers for the purposes of risk classification and the calibrating of premiums to risk status, especially in connection with the issuance of individual health insurance policies. This could lead to more applicants for such policies being designated high risk by insurers based on these tests. This in turn could increase the total number of those lacking health insurance in this country, and thus increase the number of those lacking access to adequate medical care for which insurance is the typically necessary condition.

From the libertarian-oriented commercial insurers' position, an increase in the number of uninsured in the United States may be conceded to be a misfortune. But it is not a matter of social injustice that the government is morally bound to rectify. Egalitarians disagree. The ongoing debate concerning the restructuring of health insurance in the United States, and the attendant debate concerning the "rightful" use of genetic test results under a revised national health insurance policy, will ultimately reflect a tension within each of us between simultaneously held libertarian and egalitarian commitments.

Notes

1. Kathleen Nolan and Sara Swenson, "New Tools, New Dilemmas: Genetic Frontiers," *Hastings Center Report* 18, no. 5 (October-November 1988): 41.

2. Only 15 to 20 percent of persons carrying health insurance have individually written policies. U.S. Congress, Office of Technology Assessment, *Medical Testing and Health Insurance* (Washington, D.C.: Government Printing Office, 1988), 3. Under large group policies (e.g., where the workforce of a large employer is the insured party), insurers typically do not require individual members of the group to undergo medical exams or complete medical histories when the contract is issued. "Rather, the entire group is underwritten according to factors such as the number of employees, age and gender distribution, area of the country, and prior health care costs for the entire group." Robert Pokorski, "Use of Genetic Information by Private Insurers—Genetic Advances: The Perspective of an Insurance Medical Director," *Journal of Insurance Medicine* 24 (1992): 63. Evaluation of individual risk profiles by insurers is more common in underwriting small group policies, and a number of commercial carriers do some type of risk analysis of individuals who are late applicants for coverage (e.g., as "new hires" in a workforce) under an existing

large group policy. Office of Technology Assessment, *Medical Testing and Health Insurance*, 9.

3. Currently, approximately 3 percent of applicants for individual health insurance policies are denied coverage. See Pokorski, "Use of Genetic Information," 66. The Office of Technology Assessment reports that 8 percent of applicants for individual health insurance were denied coverage. U.S. Congress, Office of Technology Assessment, *Genetic Monitoring and Screening in the Workplace* (Washington, D.C.: Government Printing Office, 1990), 62.

4. It is important not to exaggerate the "contribution" that insurers' use of genetic test results for underwriting purposes might make to the problem of inequitable access to health care in the United States. The standard practice in this country of tying health insurance to employment is clearly a major contributor to this vexing social problem.

5. Pokorski, "Use of Genetic Information," 65–67; Kenneth Abraham, *Distributing Risk* (New Haven, Conn.: Yale University Press, 1962), chap. 4.

6. Norman Daniels, "Insurability and the HIV Epidemic: Ethical Issues in Underwriting," *Milbank Quarterly* 68 (1990): 497–525.

7. Barbara Lautzenheiser, "Socialized Insurance: The Rising Tide," *Best's Review* Life and Health section, January 1989, 24.

8. Detailed explications and defenses of libertarianism can be found in Robert Nozick, *Anarchy, State, and Utopia* (New York: Basic Books, 1974); Tibor Machan, *Individuals and Their Rights* (LaSalle, Ill.: Open Court, 1989); John Hospers, "The Libertarian Manifesto," in *Morality in Practice*, ed. James Sterba, 2d ed. (Belmont, Calif.: Wadsworth, 1988), 24–33; H. Tristram Engelhart, *The Foundations of Bioethics* (London: Oxford University Press, 1986).

9. Nozick, *Anarchy, State, and Utopia*, 225.

10. Ibid., 226.

11. Material for this section draws principally on John Rawls, *A Theory of Justice* (Cambridge: Harvard University Press, 1971); Norman Daniels, *Just Health Care* (New York: Cambridge University Press, 1985); Daniels, "Insurability and the HIV Epidemic," 497–525.

12. The phrase is borrowed from Daniels, *Just Health Care.*

13. Rawls, *A Theory of Justice.*

14. Daniels, *Just Health Care.*

15. Donald Light comments: "Risk rating creates, quite deliberately, medical poverty. . . . Risk-rating sets the terms of impoverishment for each risk group. Some have low-grade impoverishment through higher premiums. Others watch their life savings depleted for a year during a waiting period and then suddenly have it stop as their bills are picked up thereafter. Still others walk over a field of manholes covered with twigs and grass, known as medical exclusion clauses. They step into one and plunge into debt." Donald Light,

"The Ethics of Corporate Health Insurance," *Business and Professional Ethics Journal* 10 (1991): 57–58.

16. Paul Menzel has cogently argued that egalitarianism's commitment to "equal access to adequate medical care" does *not* require what Menzel calls "medical egalitarianism." This is the view that "when the poor's medical needs are equal to those of others, their care should be equal." Thus, whatever medical care is available to anyone in society (e.g., to the most wealthy or perhaps the middle class) must be made available to everyone, including the poor. Menzel argues that the notion of an "adequate" or "decent" level of medical care (the level that ought to be made available to everyone, irrespective of ability to pay) is morally *compatible* with different levels of medical care available in society varying with people's willingness to pay, and morally *required* by "respect for poor persons' own rational preferences. If poor, I will certainly prefer to spend less on preserving health and saving life than if I am well off, even if in either case I am perfectly knowledgeable and rational. Lower income people will quite properly choose differently when it comes to making use of statistically expensive and marginally beneficial procedures. To flatten out these differences through uniform health care services *without* changing the basic distribution of income rides roughshod over people's preferences for the respective lives they have to live." Paul Menzel, "Equality, Autonomy, and Efficiency: What Health Care System Should We Have?" *Journal of Medicine and Philosophy* 17 (1992): 39–40.

17. Mark Pauly, "Risk Variation and Fallback Insurers in Universal Coverage Insurance Plans," *Inquiry* 29 (1992): 142–43.

[10]

Persistence and Continuity in Human Genetics and Social Stratification

Troy Duster

THE UNITED STATES IS currently in the throes of its *second* great economic transformation. As a nation, we have been here before, experiencing the wrenching disruption of old ways of life: the displacement of millions of people from their routines, their jobs, and their sense of purpose in life; the severing of old relationships; the ever increasing proportion of the population caught in aimless wandering, homelessness, crime, even a threatened sense of personal safety. For all of this, again, there is nothing new.

A century ago, the *first* great transformation was in full swing; the shift from a mainly rural, agrarian society to a mainly urban, industrial society was taking place. Commentators issued dire warnings about the end of Western civilization. They worried about the new immigrants and the industrial pollution. Their meaning was both literal, with the smokestacks, and metaphoric with the new immigrants with those who did not speak English, those with foreign tongues and alien ways. Ultimately, the early alarmists tried to explain the developments by reference to "higher and lower forms" of human life and human culture. The lower forms needed to be controlled, reduced in number, even eliminated if possible. The new science of genetics, deeply embedded in Darwinian evolutionary theory, provided convenient "answers" to the social problems of poverty and crime and mental illness that were evident in this period (1880–1925) of the first great transformation.

Today, we are seeing a parallel development. We are seeing a shift of enormous proportions as we move into our *second* great economic and social transformation, from an industrial and urban society to a postindustrial and suburban society.

As in the earlier period, important scientific breakthroughs and developments reverberate into our understandings about what it means to be human and to be flawed as well. During the last decade, the pages of scientific and popular literature have propagated the "gene myth" by printing articles that heavily imply overly simplistic "genetic solutions" to such social issues as alcoholism, mental illness, crime, and even homelessness and unemployment.[1]

However, some scholars have chosen to emphasize the sharp differences with the past when it comes to the eugenic danger. For example, the esteemed historian of science Daniel Kevles has argued that in today's society, vulnerable and marginalized groups have greater access to strategies and mechanisms for fending off any eugenic resurgence. Despite this, there is a persistent search for hard data, for a biological or biochemical explanation of homelessness, mental retardation and mental illness, alcoholism and drug abuse, "intelligence," even unemployment, crime, and violent and abusive behavior. The change has come in a thick disguise of the old concerns, the promise of untold health benefits and the lessening of human suffering. Some of these promises will be fulfilled, and some remarkable strides have already been made in the detection of genetic disorders. Yet these successes have had an unwitting and inadvertent side effect. The demonstrable advances in biomedicine have produced a halo over a host of problematic claims about the connection between genes and behavior, so much so that we are witnessing the chameleonlike reincarnation of some of the more regressive formulations of late nineteenth-century thought.

On the surface, the Human Genome Project (HGP), a program to map and sequence the entire spectrum of human genes, has as its supporting rationale the improvement of human health. The strategy is the uncovering of genetic disorders and susceptibility to disorders, and ultimately, the hoped-for development of gene therapies to treat or cure those disorders.[1] But no matter how one slices it, just underneath the talk of a paradigm shift, we as a society seem inexorably pulled back to the ancient concern for what could be called trouble at the bottom, virtue at the top. The fact that the banner of health, medicine, and science waves over the new biotechnologies has lulled us into a complacency, even a receptivity, to a rehearing of claims made a century ago when the science of human genetics, in its infancy, was most seductive.

From Biological Darwinism to Social Darwinism

One of the most enduring truths in the study of human social life is that all societies are stratified. The unequal access to valuable resources can be based on something as simple as age or as complex as claims to spiritual or intellectual power. But as far back as recorded history permits us to garner evidence, we also know that human beings have always tried to justify that stratification. In *The Republic*, Plato creates the "myth of the metals" to justify why only those "born gold" can become philosopher kings. The notion that power and privilege are *inherited* has a longer and wider history than the notion that power and privilege are *achieved*. The link between a theory of human biology and social theory has always been a significant force in the history of ideas, but only in the last 150 years has the connection donned scientific clothing. At the core of this relatively recent development is the direct link between biological Darwinism and social Darwinism, and the direct but underappreciated implications for the birth of human genetics. To appreciate the subtle, sometimes subterranean continuity between the past and the present, we must go back to those early beginnings.

Charles Darwin's *Origin of Species* is the bible of evolutionary theory, at once a meticulous classification system of organisms and a theory of the evolving relationships between them. In its simplest form, the implications of Darwinian taxonomy are known even to some grade-school children: at the bottom of the rung is the single-celled amoeba and at the top of the heap is the magnificently complex human. In between are all the combinations and permutations and mutations that form an intricate hierarchy of organisms. It is intricate. It is most decidedly a hierarchy.[2]

What of human beings? Once we get to the top rung of the ladder of species evolution, biological Darwinism trails off and, like a relay sprinter in a race, huffing and puffing and tired, hands the baton onto the runner in the next leg. The baton is passed from biological Darwinism to social Darwinism.

In the biological version of adaptation, species are ranked along a hierarchy of complexity in evolutionary adaptation, but what about rankings *within species*? Within, between, and among human groups, was there not also an evolutionary tree? As Darwin had done for biological Darwinism, the English social theorist Herbert Spencer (1820–1903) would issue the canon of social Darwinism. To convey a better understanding of the climate in which scientific genetics germinated, it is necessary to rescue and restate two important features of late nineteenth-century thought that have been largely forgotten. The first is that Spencer dominated the social thought of his age as few have ever done. He was by far the most popular nonfiction writer of his era; his ideas were so popular that he sold more than 400,000 copies of his books during his lifetime.[3] In the United States, by the turn of century Spencer attained the status of a dominant cultural figure among a wide range of American politicians, intellectuals, educators, and public policy advocates.[4] Indeed, he was so influential that Oliver Wendell Holmes once sardonically turned to his colleagues on the Supreme Court to remind them that "Herbert Spencer did not write the U.S. Constitution."[5]

Although Darwin would ultimately distance himself from the more regressive social implications of Spencer's social evolutionary theory, Darwin once called Spencer "about a dozen times my superior."[6] Perhaps more significantly Francis Galton (1822–1911), the man who coined the term "eugenics" and who was the founder of the eugenics movement, deeply admired Spencer, and was influenced by Spencer's thinking on social evolution.[7] Not so coincidentally, Galton was also the father of human population genetics, a statistical method designed to study patterns of inheritance as a way of explaining evolutionary developmental stages *within and between humans.*

Second, although Charles Darwin set the stage, it was Spencer, not Darwin, who would develop the key concepts that would apply evolutionary theory to human beings. It was Spencer, for example, who coined the phrase "the survival of the fittest."[8] Herbert Spencer focused his ideas not on the animal kingdom, but on social life, human behavior, and the internal *differences of evolution among humans.*

Spencer's influence upon a newly emerging field of anthropology, the "study of man," was also overwhelming.[9] Not only are human beings to be arrayed along a continuum of evolutionary development, but so are the races and the cultures, societies, tribes, and nations in which they live. At an individual level, the idea of a "savage" or a "primitive" was at one end of that continuum, and at the other was the "civilized person." So, too, there was the notion of a primitive or savage society.

The fundamental basis of the continuum from savage to civilized, wrote Spencer, was the developmental stage of the brain. That was to be explained, in turn, by the way in which humans adapted to nature, and in particular, the seasons and the passage of time. The "primitive peoples" only had a sense of time relevant to such natural events as when birds migrate, or when fall or winter or spring begins. The more advanced and more "civilized" could encompass decades, even centuries into their thinking, planning, and "accumulation." As such, their brain capacity was vitally stimulated and literally enlarged. The longer the time sequence a human could encompass, said Spencer, the higher the level of intellectual development. At the bottom of the heap were the Australian Aborigines. Just above them were the Hottentots, who were judged one notch superior because they could use a combination of astrological and terrestrial phenomena to make adjustments to time sequences and changes.[10] Moving on up, the next on Spencer's social evolutionary ladder were the nomads, just a rung below the settled primitives who lived in thatches and huts. Since they stored goods for future use, their conception of and relation to time were "more developed."

Anthropology, the new scientific study of human groups across all human societies, was born in this same period of evolutionary theory and was saturated by it. Just as human beings can be stratified according to their social evolutionary developments, it was argued, so can their cultures. It followed that once selected individuals from "inferior cultures" came to live in "superior cultures," there would be a limit as to what their brains, of lower development capacity, could handle. Writing exactly a century before this claim would be made again by Arthur Jensen,[11] Spencer noted in 1869 that black children in the United States could not keep up with whites because of the former's biological and genetically endowed limits, "their [Blacks'] intellects being apparently incapable of being cultured beyond a particular point."[12]

This reached its logical culmination in the work of James George Frazer (1854–1941), who produced the prodigious six volumes of *The Golden Bough* that formally stratified cultures and societies along a continuum from simple to complex, from savage to civilized. Frazer posited a three-stage hierarchical theory, that human societies evolve from magic to religion and, finally, to science.[13] At the bottom of the hierarchy, of course, were "primitive cultures."

It is well known that Darwin's biological evolutionary theory was not accepted among the Christian clerics at the time.[14] Equally important for the birth of nineteenth-century scientific human genetics, the church had fiercely contested the biological evolutionary theory of the ladder from lower creatures to humans, and the church also had a strong vested interest in attacking the stage theory of social Darwinism. The idea that, over time, humans ascend to higher and higher forms of linguistic complexity, moral reasoning, and spiritual existence was counter to prevailing Christian theology. Four years before Darwin published the *Origin of Species*, Bishop Richard Whately published *On the Origin of Civilization*. Whately invoked a modern-day version of sociolinguistics to provide empirical evidence that the human race declined, not improved, over the millennia.[15] In the late eighteenth century, scholars had turned their attention to the origins of European languages. A body of work had developed indicating that there was probably an original common tongue that mothered Sanskrit, several Indo-European languages, and some Asiatic languages.

This theory resonated with the idea that a falling away from basic religious truths was the fate of humanity.[16] A fundamental tenet of Christian theology was that humans had declined, not ascended, over time.[17] We must recall that Christianity posited an early state of perfection and a "fall from grace." Thus, it was not an uncontested position to argue that, in the beginning, there had been savagery and primitivism and terrible warfare and sacrifices. To place science at the apex of evolutionary developments of human cultures was at odds with a strong strain of thought among theologians. There was thus popular and clerical resistance to these ideas. Many groups had to be persuaded. Enter social Darwinism.

That was the social and political context reflected in James Frazer's *The Golden Bough*. We can now see some of the reasons for the author's fervid assertions of counterevidence that "primitive man" was an early, barbaric, savage being. Wittgenstein understood this well. He cast suspicion on Frazer's motives for recounting stories of how terrible things are among "primitive peoples" in the following way: "Frazer [tells] the story . . . in a tone which shows that something strange and terrible is happening here. And that is the answer to the question 'why is this happening?': because it is terrible."[18]

Wittgenstein was hardly alone in his attack on the tautological and self-serving presentation of data by Frazer. The eminent French social theorist Emile Durkheim also dissected the key architecture of Frazer's argument. Durkheim held that there was nothing "primitive" about totemic organization among the so-called primitives, but that instead, this served complex and parallel social functions with what occurs in "advanced" societies.[19]

The search for empirical evidence to document social evolution within *Homo sapiens* is the subject of a full monograph fully documented in Stephen Gould's *The Mismeasure of Man*.[20] Gould reveals the painstaking care with which nineteenth-century scientists sought to "prove" that the size of the skull could be arrayed along an evolutionary continuum, with white males at the apex. When they failed, they improvised, or compromised, or "finessed" the data.

Again, in periods of great social, economic, and political turmoil, there is a special appeal of genetic explanations of stratification to privileged persons. The conflation of privilege, science, and ideology is apparent in the life and work of Charles Davenport,[21] arguably the most influential American human geneticist of his day. Davenport was also a eugenicist, who fused genetics and eugenics in ways that are, in retrospect, embarrassing. Most of the enthusiasts for human eugenics, then as now, were not formally trained practicing geneticists. Davenport, however, was a highly esteemed biological scientist and an unreconstructed Mendelian, who enthusiastically embraced eugenics. He went so far as to count up the number of male offspring from his Harvard graduating class (141), and when he contrasted that with the number of males in his class (278), he sounded the following alarm:

> Assuming that a class matures half as many sons as it graduates and that their descendants do the same for six generations, 1000 Harvard graduates in the 1880s will have sixteen male descendants of the 2080s. These sixteen sons will be ruled by the scores of thousands of descendants of 1000 of the Rumanians, Bulgarians,

Greeks and hybrid Portuguese of the 1880s. Such figures must make one fear for the future.[22]

What Davenport feared was that these "lower forms of human life" on the evolutionary scale would someday rule the "higher forms" from the old Harvard stock. In its crudest formulation, as we shall see, this idea would take an ugly political and public policy turn, both the strong advocacy and the practice of forced sterilization of those at the bottom of the social order.

> The lowest stratum of society has, on the other hand, neither intelligence nor self-control enough to justify the State to leave its matings in their own hands. On the contrary, the defectives and the criminalistic are, so far as may be possible, to be segregated under the care of the State during the reproductive period or otherwise forcibly prevented from procreation.[23]

Ronald Fisher, generally credited appropriately as the founder of modern statistics, invented most of the statistics (between 1916 and 1932) now used in biological research, including the *analysis of variance*. Fisher was the author of a very influential book entitled *The Genetical Theory of Natural Selection* in which he proposed a theory of evolution. Published in 1930, the book sounded the familiar alarm (by that time) that since the lower classes were outbreeding the upper classes, the human species was in danger of deterioration.[24] Fisher's statistical genius created a détente, if not a peaceful resolution, between population and Mendelian genetics. With the analysis of variance he developed a statistical strategy for assigning different weights to different causes, most particularly relevant, between "heredity" and "environment." But Fisher also used statistics to buttress and embellish Mendelian genetics: "Simultaneously he was the inventor of the modern theory of quantitative genetics that shows how Mendel's laws can be used to generate the observed similarity between relatives in continuously varying traits like height, weight, and IQ scores."[25]

To better appreciate the continuity with the past, we must step back and place this development within the context of the great transformation in the social, political, and economic circumstances of the period.[26] Then, as now, the societies in which the appeal was so great were themselves undergoing the first of two great social transformations.

The first of these two social and economic tranformations took place at the turn of the century. Hundreds of thousands of Europeans were immigrating to the United States annually. During the three decades from 1885 to the First World War, more than 15 million Europeans came to the U.S. shore as permanent émigrés.[27] Prior to 1885, European immigration had been primarily from England, Scandinavia, and Germany. However, in the thirty-year period from

1885 to the First World War, the "new immigrants" were predominantly Italian, Russian, Polish, Slavic, and Jewish (the latter cutting across several national boundaries). The already-arrived (English, Scandinavian, and German Americans) saw real threat in the new immigrants.[28] Across many spheres of American life, the self-proclaimed "full-blooded" northern and western European Americans pooled their resources and moved collectively to block further "alien" immigration. Ultimately, they would succeed. Congressional action in 1924 would slam the door almost completely shut on people from the southern, central, and eastern parts of Europe.

For example, yearly immigration from the Baltic states and Russia was 250,000 in both 1913 and 1914, but dropped to 21,000 in 1923, then dropped to 1,000 per year for the next fifty years.[29] Similarly, Italian immigration dropped from 222,000 in 1921 to 56,000 in 1924, dropped even more in the next decade, and never again approached the 1924 level. In congressional testimony to justify the new legislation to stop the flow of this "lower form of human life," the "scientific" IQ test became a powerful justification device. Jews (and Russians and Italians) had been revealed by the Binet test as genetically inferior, more prone, biologically, to "feeblemindedness." "The mental testers pressed upon the Congress scientific IQ data to demonstrate that the 'New Immigration' from southeastern Europe was genetically inferior. That contribution permanently transformed American society."[30]

The United States passed immigration laws in the 1920s that were unashamedly committed to racial and ethnic quotas, and nearly a half century was to elapse before they would be substantially repealed. The first restrictive legislation in 1921 set quotas at 3 percent for immigrants from any nation then resident in the United States. In the next two years, Congress was besieged by eugenicists' testimony, arguing from IQ test data collected by the army during universal conscription from the First World War. Just closing the borders of the nation was not enough. The eugenicists wanted to impose harsher quotas on the nations of southern and eastern Europe because the test scores were used to show that Italians, Slavs, Jews, and Poles came from inferior racial stock.[31] They won the fight, and a new, more restrictive law was passed in 1924, resetting the quotas at 2 percent from each nation recorded in the 1890 census. Since northern Europeans and the British had predominated in 1890, this effectively shut off the "flood of immigration" from southern, south central, and eastern Europe.

This was part of the first great social and economic transformation of the United States from agrarian to industrial and urban society. The accompanying social problems and social troubles made for a receptive audience, among both scientists and laypersons, that the problems of poverty, crime, mental retardation, and mental illness could best be explained by reference to the qualities of the individuals who brought the problems with them, the "lower forms on the evolutionary tree" or from "lower forms of cultural life." Even the infant mortality rate was explained by reference to "qualities inherent" to those so victimized:

> The negro infant death rate is in every district higher than the white rate. Throughout the Austro-Hungarian and Russian districts, with very high density of population and great poverty, the infant mortality is exceptionally low. There can be no question that the low rate is due to the qualities inherent in the people themselves.[32]

As the United States and other industrialized nations now move into their second great transformation—from industrial sector– to service sector–dominated societies—we are witnessing a similar set of social problems of high levels of endemic unemployment, high rates of homelessness, social dislocations and the sense of less safe streets, and the like. As in our past, we swing the pendulum back and forth between explaining these developments by reference to the qualities and characteristics of individuals (and the societies or cultures from whence they come) versus the social and economic forces that might also help explain these dislocations. In this swing of the pendulum, human genetics has regularly played an important role.

Genetic Reductionism: The Sociohistorical Continuity

Although many are aware of the gross human rights abuses in the name of eugenics[33] of the early part of the century, most of the current advocates, researchers, and celebrants of the putative link between genetic accounts and socially undesirable behavior (or characteristics or attributes) are unaware of *the social context of that history,* or they are too quick to dismiss that history as something that happened among the unenlightened. Both formulations miss the special appeal of genetic explanations and eugenic solutions to the most privileged strata of society. Well into the middle part of this century, bankers and politicians, governors, university professors, the president of Stanford University, and other respected professionals favored the sterilization of the "lower forms" of human life.

Every era is certain of its facts. The heyday of the eugenics movement was no exception, sure that feeblemindedness, degeneracy, and criminality were inherited. In 1912, the American Breeder's Association, an organization of farmers and university-based theoreticians, created a Committee to Study and to Report on the Best Practical Means of Cutting Off the Defective Germ Plasma in the American Population. It was a five-man committee, chaired by a prominent New York attorney, and having among its membership a prominent physician from the faculty at Johns Hopkins. At the 1913 meeting of the association, the report was delivered[34] and read in part:

> Biologists tell us that whether of wholly defective inheritance or because of an in-surmountable tendency toward defect, which is innate, members of [select] classes must generally be considered as socially unfit and their supply should if possible be eliminated from the human stock if we would maintain or raise the level of quality essential to the progress of the nation and our race.

Sterilization became the method of choice for eliminating the socially unfit. California had one of the longest-running involuntary sterilization programs in the country. In 1927, a team of prominent and respected citizens was assembled to consult on the effectiveness of this program. Participants included Lewis Terman, the most prominent psychometrician in the country, David Starr Jordan (president at Stanford), and S. J. Holmes, a distinguished geneticist from Berkeley. Covering the period from 1909 to 1927, a series of reports came out of the group that produced "the first comprehensive 'proof' that sterilization was cost-effective and posed no significant medical harm to the institutionalized persons at whom it was aimed."[35]

In the 1920s, two major legal developments would shape and be shaped in turn by increasingly dominant views of "race betterment" through biological science and its applications. The first would be new immigration laws, strongly backed by the old American stock, to close off the immigration doors to those from southern and eastern Europe. It is well known and fully documented how geneticists and eugenicists provided vital testimony before the U.S. Congress in the early 1920s.[36] This resulted in the passage of two laws, which in effect reduced immigration from a surging high of approximately half a million a year to less than ten thousand by the end of that decade. The strategy was overwhelmingly popular among local politicians in several key states. For example, the Virginia legislature passed an involuntary sterilization bill (30-0 in the Senate, 75-2 in the House), noting that "heredity plays an important part in the transmission of insanity, idiocy, imbecility, epilepsy and crime." That law gave the power to superintendents of five state institutions to petition for permission to sterilize selected inmates.

On May 2, 1927, the Supreme Court upheld Virginia's involuntary sterilization law, opening up not only a floodgate of sterilizations in the United States, but a model that would soon be adopted, expanded, and forever made infamous by Hitler's Third Reich. In mid-July 1933, Germany enacted a eugenic sterilization law. The U.S. eugenicists provided the intellectual and ideological underpinnings and were widely cited as the genetic authorities in behalf of this development. California was one of the leading states in the country in terms of its use of involuntary sterilization laws. From 1930 to 1944, more than 11,000 Californians were sterilized under these laws. The Germans cited the California development as a model (in 1936, Heidelberg University awarded honorary degrees to several key U.S. eugenicists), but they took it much farther. In the first year of the German program, 52,000 were placed under final order to be sterilized, a development that was in turn hailed by U.S. eugenicists. In the period from 1933 to 1945, the best estimates indicate that the Nazis sterilized approximately 3,500,000 people.[37]

Genetic Explanations of Violence and Deviance Today

Today, the United States is heading down a similar road with an unfounded faith in the connection between genes and social outcomes. This is being played out on a stage with converging preoccupations and tangled webs that interlace crime and violence, race, and genetic explanations. So-called genetic studies of criminality have a heavy dependency on incarcerated populations. Thus, for example, one of the more controversial issues in the "genetics" of crime is whether males with the extra Y chromosome, or XYY males, are more likely to be found in prisons than are XY males.

The first major study suggesting a genetic link came from Edinburgh, Scotland. While all of the 197 males in the account of prison hospital inmates were described as "dangerously violent,"[38] seven had the XYY karyotype. The seven males constituted about 3.5 percent of the total. But since estimates are that only about 1.3 percent of all males have the XYY chromosomal makeup, the authors posited that the extra Y significantly increased one's chances of being incarcerated. Ever since, a controversy has raged as to the meaning of these findings and the methodology that produced them. Notice the logic of such studies: they argue the genetic link to crime, but rely primarily upon incarcerated populations. Yet incarceration rates are a function of incarceration decisions, a fact that social science research has long shown to be a function of social, economic, and political factors.[39]

In the last decade, the United States has been building more prisons and incarcerating more people than at any other moment in our history. Indeed, in the brief period from 1981 to 1991, we went from a prison population of 330,000 inmates in state and federal prisons to 804,000. That rate constitutes substantially more than a doubling in a single decade, *the greatest rise in a prison population in modern history.*[40]

Converging on this development is the racial patterning of those arrested and serving time. African Americans are incarcerated at a rate approximately seven times greater than that of Americans of European descent. While the current incarceration rate of African Americans in American prisons is approximately seven times that of white Americans, this is a very recent development.

Although the XYY controversy was not about race, the logic and reasoning behind this kind of study create a foreboding atmosphere for a racial theory of violence, crime, or homicide. The method of the XYY study was to compare the characteristics of those incarcerated with those outside prison. If the characteristic under investigation (in this case, the extra Y chromosome) was more prevalent, then it was assumed that there must be a "genetic" reason for it. Notice that since the rate of incarceration for African Americans is seven times that of whites, this logic would produce a "genetic" explanation for the higher rate of incarceration for African Americans. The XYY data have long since been shown to be insubstantial and problematic. However, lest anyone have the impression that these are theoretical issues or debatable points that do not have consequences in practice:

> In a few score of reported cases, an XYY fetus has been unexpectedly diagnosed during prenatal genetic studies that were initiated because of advanced maternal age or other reasons. . . . In the United States, an informal survey by one of the leading researchers on XYY males, Dr. Arthur Robinson of Denver, showed that about 50 per cent of parents elected to terminate such pregnancies.[41]

If we are ignorant of recent history and do not know that the incarceration rate and the coloring of our prisons are a function of dramatic changes in the last half century, we would be far more vulnerable to the seduction of a genetic explanation.[42] A review of prison incarceration rates by race for the last half century reveals an astonishing pattern that should give pause to anyone who would try to explain the developments by reference to the genetic makeup of the incarcerated. The gene pool among human beings takes many centuries to change, but since 1933, the incarceration rates of African Americans in relation to whites have gone up in a striking manner. In 1933, African Americans were incarcerated at a rate of approximately three times the rate of incarceration for whites. In 1950, the ratio had increased to approximately 4 to 1; in 1960, it was 5 to 1; in 1970, it was 6 to 1; and in 1989, it was 7 to 1.[43]

Today's celebrants of the new advances in human genetics conjure up a future of the elimination of genetic disorders, either by prevention of the births of those who have such disorders or by the promissory note of cure and treatment. Yet while we can detect and diagnose hundreds of genetic disorders, effective cures are rare.[44] Thus, it is to prevention that we are bound to turn.

The Violence Initiative and the Medicalization of Deviance

In September 1991, the National Institute of Mental Health issued a program announcement[45] with the title *Research on Perpetrators of Violence*. The announcement was explicit in its aim to further research on cost-efficient measures, beginning with clinical assessment, that might "prevent" violence:

> The purpose of this announcement is to encourage investigator initiated research on the etiology, course, and correlates of aggressive and violent behaviors in children, adolescents, and adults. Through this announcement, the National Institute of Mental Health (NIMH) expects to support research that will improve the scientific base for more effective and cost-efficient approaches to clinical assessment, treatment, management and prevention.

In specifying the priority areas of research interest, the Violence and Traumatic Stress Research Branch specified that "studies may focus on risk factors and procedures that contribute to the occurrence and influence the course of violent behaviors (e.g., neurochemicals and neuroendocrines, parent-child rearing practices . . .)." Here is the summary language of one of the funded project's major aims:

> This research seeks to contribute to the understanding of the origin of serious and chronic aggressive and antisocial behaviors so that preventive interventions can be more effectively timed and targeted. . . . The project involves a longitudinal design in which 310 parent-infant male dyads from low SES backgrounds will be assessed when the child is 18, 24, and 42 months of age.[46]

Yet another funded project describes its method of selection as follows: "One-hundred and twenty Black third-grade boys will be recruited from 10 classrooms in the Durham, North Carolina school system. Dyad type will be assessed by pupil and teacher ratings of the extent to which pairs of boys in the classroom initiate aggression against one another."[47]

After the urban civil disturbances in Los Angeles triggered by the verdict handed down in the Rodney King case, James Q. Wilson wrote an opinion piece for the *Wall Street Journal* that was widely distributed in a syndicated column around the nation.[48] At the conclusion of this article, Wilson strongly suggested that the high rate of black crime, six to eight times higher than that of white crime, may be reduced only by profound interventions on the lives of black male infants from birth to age ten:

> The best way to reduce racism real or imagined is to reduce the black crime rate to equal the white crime rate. . . . To do this may require changing, in far more profound and all-encompassing ways than anything we now contemplate, the lives of black infants, especially boys, from birth to age 8 or 10. We have not yet begun to think seriously about this, and perhaps never will. Those who must think about it the hardest are those decent black people who must accept, and ideally should develop and run, whatever is done.[49]

Intervene in "profound and all-encompassing ways"? "Develop and run" a program to deal with these young black infants and children? If these same words were written by someone with a long history of emphasizing environmental explanations of criminal behavior, it would be one thing. But James Q. Wilson teamed up with Richard Herrnstein in 1985 to publish a book, *Crime and Human Nature*, that explicitly argued that U.S. society take a more serious look at biogenetic explanations of crime. Indeed, in June 1992, James Q. Wilson published an essay in which he stated,

> We know now that the child brings a great deal to the parent-child relationship, that many aspects of personality have genetic origins, and that some infants experience insults and traumas. . . . Two children in the same family often turn out very differently. This casts great doubt on the notion that the shared environment of the children is a principal—or even a very important—factor in their development. What is going on here?[50]

Indeed, what is going on here? Having cast doubt that environment may be even given the status of "important," Professor Wilson then goes on to offer as one of three hypotheses something that sounds very much like the Spencer of 1869 and the Jensen of 1969: "That children with low IQ's find school work boring and frustrating and turn to physical activity—including rowdy, violent activity—as an alternative source of rewards."[51]

In the essay immediately after the Rodney King verdict was announced and the urban uprisings occurred, James Q. Wilson said that a program more profoundly different "than anything we now contemplate" should "*ideally*" be in the hands of "decent black people." But then, since we do not live in an ideal world, we may have to settle for someone else running "these programs"—perhaps others than the "decent black people who must accept . . . whatever is done."

The deep and recurring concern with violent crime is still very much with us. The link between theories of crime, biology, and violence is still with us, and it still fuels thought at the highest levels of government, medical, and biological research. In February 1992, Frederick Goodwin, a top official in the Bush administration, made some remarks at a meeting of the National Mental Health Advisory Council. Dr. Goodwin was at the time the chief administrative officer of three federal agencies coordinated under his jurisdiction: the National Institute for Alcohol Abuse Administration, the National Institute on Drug Abuse, and the National Institute of Mental Health. The consortium is conjoined as the Alcohol, Drug Abuse, Mental Health Administration, or ADAMHA. As then chief of ADAMHA, Dr. Goodwin made remarks at that meeting of February 11, 1992, that would get him "demoted" to being head of only one of the agencies, the National Institute of Mental Health.

Goodwin has said that his remarks were taken out of context. But what was the proper context of these remarks?

> Somebody gave me some data recently that puts this in a perspective and I say this with the realization that it might be easily misunderstood, and that is, if you look at other primates in nature—male primates in nature—you find that even with our violent society we are doing very well.
>
> If you look, for example, at male monkeys, especially in the wild, roughly half of them survive to adulthood. The other half die by violence. That is the natural way of it for males, to knock each other off and, in fact, there are some interesting evolutionary implications of that because the same hyper-aggressive monkeys who kill each other are also hyper-sexual, so they copulate more and therefore they reproduce more to offset the fact that half of them are dying.
>
> Now, one could say that if some of the loss of social structure in this society, and particularly within the high impact inner city areas, has removed some of the civilizing evolutionary things that we have built up and that maybe it isn't just the careless use of the word when people call certain areas of certain cities jungles, that we may have gone back to what might be more natural, without all of the social controls that we have imposed upon ourselves as a civilization over thousands of years in our own evolution.
>
> This just reminds us that, although we look at individual factors and we look at biological differences and we look at genetic differences, the loss of structure in society is probably why we are dealing with this issue and why we are seeing the doubling incidence of violence among the young over the last 20 years.[52]

His statement is infused with social evolutionary theory, and Goodwin explicitly links the concerns for biological and genetic differences to the social milieu. When reflecting upon the history of the link between human genetics and eugenics, we may be tempted to suggest that only the fringes and the marginal actors were involved. That is not true. Charles Davenport was among the most influential and respected geneticists of his time, and he was an ardent eugenicist. I am not suggesting that Goodwin is a eugenicist. He is certainly not at the fringes of biomedical research on the causes of violence. While he was director of the National Institute of Mental Health, he was a strong advocate of a program of research to try to better explain the "violent, junglelike" behavior of youth analogized to "hyper-aggressive monkeys." The fact that he invoked evolutionary theory to explain a potential genetic link to a proneness to violence is a reminder that we are our past.

Conclusion

The United States has experienced two great social and economic upheavals that have produced massive transformations in the society. The first was the shift from rural agrarian life to urban industrial life, often called the Industrial Revolution and centered at the turn of the last century. Massive dislocations of people occurred, producing internal migration from rural sites to cities, and internationally across the oceans and seas and national boundaries. These disruptions in the life courses of scores of millions of people were accompanied by homelessness, poverty, crime, violence, and a strong sense of insecurity and foreboding about those at the bottom of the social and economic order.

In this setting, the idea that social stratification and the behavior of those at the bottom of the economic order are biologically and genetically inferior was seductive enough to alter public policy, law, and the "medical" practice of forced sterilization. The blurring of scientific work in genetics and the social policy advocacy of human eugenics (the "betterment" of the human race cast in biological terms) was sufficient to produce congressional testimony and, ultimately, restrictive immigration aimed at securing the national border of the country against "the genetically inferior."

A century later, the nation is currently in the throes of another great economic and social transformation. We are no longer the predominantly urban and industrial society we were at midcentury; rather we are witnessing dramatic shifts in the population composition, structure, and functioning of our greatest urban centers. Meanwhile more than three-quarters of the nation's employment is in the service sector. As in the late nineteenth century, this late-twentieth-century shift has produced disruptions in the lives of persons who were previously relatively secure in an industrial, near full-employment economy. The rising youth unemployment is accompanied by increased homelessness, poverty, and the same sense of personal insecurity and the concerns of violence at the bottom of the social order. And once again, the seductive voice of a biological and genetic account is on the horizon.

A review of the *Reader's Guide to Periodical Literature* from 1976 to 1982 revealed a 231 percent increase in articles that attempted to explain the genetic basis for crime, mental illness, intelligence, and alcoholism during this brief six-year period. Even more remarkably, between 1983 and 1988, articles that attributed a genetic basis to crime *more than quadrupled* during this period in their frequency of appearance rate from the previous decade. This development in the popular print media was based in part upon what was occurring in the scientific journals. During this period, hundreds of articles appeared in the scientific literature,[53] making claims about the genetic basis of several forms of social deviance and mental illness.[54]

At first glance, it would appear that this was an outgrowth of what was happening *in the field of molecular genetics.* Many important breakthroughs were occurring during this period, including the increasing ability for intrauterine detection of congenital disabilities. However, the resurgence of genetic claims has not been coming from those working at the vanguard laboratories in molecular biology or biochemistry. Indeed, given the requirements of up-to-the-minute monitoring to keep abreast of technological and scientific breakthroughs in these fields, it is of special interest to note that the major database for the resurgent claims is a heavy reliance upon Scandinavian institutional registries dating back to the early part of the century.[55] In the November 1993 issue of *American Health*[56] the cover asks, "Bad Seeds: Is Violence in Your Genes?" The header for the story entitled "Born Bad?" states unequivocally that "New Research Points to a Biological Role in Criminality." Yet, once again, the heavy reliance is upon correlations achieved relating data collected more than a half century ago in Scandinavia.

We need to stay ever alert to the subtle continuities and persistencies in the attempt to explain trouble at the bottom of the social and economic ladder by too easy answers provided by the "newest developments in biology" that grant a scientific cover of legitimacy.

Notes

1. For an analysis of the "gene myth" see Ted Peters, *Playing God: Genetic Determinism and Human Freedom* (New York: Routledge, 1997).
2. A specific promise of such hope is offered by C. Thomas Caskey, "Presymptomatic Diagnosis: A First Step Toward Genetic Health Care," *Science* 262 (October 1, 1993). For other expressions and comments on this new hope, see, for example, Daniel J. Kevles and Leroy Hood, *The Code of Codes: Scientific and Social Issues in the Human Genome Project* (Cambridge: Harvard University

Press, 1992); Robert N. Proctor, "Genomics and Eugenics: How Fair Is the Comparison?" in *Gene Mapping: Using Law and Ethics as Guides*, ed. George J. Annas and Sherman Elias (New York: Oxford University Press, 1992), 57–93; Robert N. Proctor, "Eugenics Among the Social Sciences: Hereditarian Thought in Germany and the United States," in *The Estate of Social Knowledge*, ed. JoAnne Brown and David K. van Keuren (Baltimore: Johns Hopkins University Press, 1991), 175–205; Jerry E. Bishop and Michael Waldholz, *Genome* (New York: Simon & Schuster, 1990).

3. There are some missing links, and indeed, as unreconstructed creationists (and not a few hard-nosed empiricists) are happy to point out, some faith.

4. Spencer actually sold 300,000 copies of his books in the United States alone. See John S. Haller Jr., *Outcasts from Evolution: Scientific Attitudes of Racial Inferiority, 1859–1900* (Urbana: University of Illinois Press, 1971), 128.

5. Richard Hofstadter, *Social Darwinism in American Thought* (Boston: Beacon, 1955).

6. Objecting to an opinion from the majority, Holmes's dissent included this statement: "The Fourteenth Amendment does not enact Mr. Herbert Spencer's Social Statistics." See William Seagle, *The History of Law* (New York: Tudor Publishing Co., 1946), 417 n.

7. This was from a letter written December 10, 1866, to Joseph Hooker. See Francis Darwin, ed., *The Life and Letters of Charles Darwin*, 2 vols. (New York: Basic Books, 1959), 2:239. Darwin would write in another letter to E. Lankester that "I suspect that hereafter he [Spencer] will be looked at as by far the greater living philosopher in England; perhaps equal to any that have lived" (2:301).

8. See Daniel J. Kevles, *In the Name of Eugenics* (Berkeley and Los Angeles: University of California Press, 1985), chap. 1.

9. Not aware of this history, current common sense and conventional wisdom unreflexively attribute this notion to Darwin, since it is now also vital to biological Darwinian theory of the evolutionary tree in the animal kingdom.

10. Herbert Spencer, *The Study of Sociology* (New York: D. Appleton and Co., 1896); Edward Burnett Tyler, *Primitive Culture* (London: John Murray, 1871). Both Spencer and Tyler offered similar treatises about the evolution of cultures, and each claimed priority and accused the other of plagiarism. Nonetheless, it is without challenge that Spencer was the more influential social thinker of the era.

11. Herbert Spencer, *Principles of Sociology*, 2 vols. (New York: D. Appleton and Co., 1899).

12. Arthur Jensen, "How Much Can We Boost IQ and Scholastic Achievement?" *Harvard Educational Review* 39, no. 1 (winter 1969): 1–123.

13. Haller, *Outcasts from Evolution*, 124.

14. Frazer did not posit a lockstep evolution. He acknowledged a zigzag path to

"progress" and liked the metaphor of waves sweeping up the shore line, ever constantly progressing even as there is receding at times.

15. Even into the 1920s with the famous Darrow trial, and even into the late twentieth century, there are still strong holdouts against evolutionary theory.

16. Robert Frazer, *The Making of the Golden Bough: The Origins and Growth of an Argument* (New York: St. Martin's Press, 1990), 13.

17. Ibid.

18. In the middle of the nineteenth century, the theory had prominent proponents who lectured at Oxford and other major universities in the West.

19. Frazer, *Making of the Golden Bough*, xiii.

20. Emile Durkheim, *Elementary Forms of the Religious Life* (London: Allen and Unwin, 1957); A. Moret and G. Davy, *From Tribe to Empire: Social Organization Among Primitives and in the Ancient East* (New York: Knopf, 1926).

21. Stephen Jay Gould, *The Mismeasure of Man* (New York: Norton, 1981).

22. Charles B. Davenport, "The Eugenics Programme and Progress in its Achievement," in *Eugenics: Twelve University Lectures* (New York: Dodd, Mead, 1914).

23. Ibid., 11.

24. Ibid., 10.

25. Richard Lewontin and Michel Schiff, *Education and Class: The Irrelevance of IQ Genetic Studies* (New York: Oxford University Press, 1986), 10.

26. Ibid.

27. This section is based upon a larger segment from my *Backdoor to Eugenics* (New York: Routledge, 1990), where this argument is more extensively developed.

28. Elbridge Sibley, "Some Demographic Clues to Stratification," in *Class Status and Power*, ed. S. M. Lipset and R. Bendix (Glencoe, Ill.: Free Press, 1953), 382.

29. Leon J. Kamin, *The Science and Politics of I.Q.* (New York: Humanities Press, 1974); Joseph Gusfield, *Symbolic Crusade* (Urbana: University of Illinois, 1963); Mark H. Haller, *Eugenics: Hereditarian Attitudes in American Thought* (New Brunswick, N.J.: Rutgers University Press, 1963); Kenneth M. Ludmerer, *Genetics and American Society* (Baltimore and London: Johns Hopkins University Press, 1972).

30. Stanley Lieberson, *A Piece of the Pie* (Berkeley: University of California Press, 1981), 9.

31. Kamin, *The Science and Politics of I.Q.*, 12.

32. Gould, *The Mismeasure of Man*, 232.

33. *Eugenical News* 1, no. 11 (November 1916): 79.

34. Daniel J. Kevles, "Controlling the Genetic Arsenal," *Wilson Quarterly* 16, no. 2 (spring 1992): 68–76.

35. H. H. Laughlin, *The Scope of the Committee's Work*, Bulletin no. 10A (Cold Spring Harbor, N.Y.: Eugenics Records Office, 1914), 1213.

36. Philip Reilly, *The Surgical Solution: A History of Involuntary Sterilization in the United States* (Baltimore: Johns Hopkins University Press, 1991).

37. Haller, *Eugenics*, and Ludmerer, *Genetics and American Society.*

38. Reilly, *The Surgical Solution.*

39. P. Jacobs, A. M. Brunton, M. Melville, R. Brittain, and W. McClemont, "Aggressive Behavior, Mental Subnormality, and the XYY Male," *Nature* 208 (1965): 1351–52.

40. Elliott Currie, *Confronting Crime: An American Challenge* (New York: Pantheon, 1985); Jerome Skolnick, *Justice Without Trial: Law Enforcement in a Democratic Society* (New York: Wiley, 1966).

41. U.S. Department of Justice, Office of Justice Programs, Bureau of Justice Statistics, NCJ-133097 (Washington, D.C.: Government Printing Office, 1992), document 1:3.

42. Aubrey Milunsky, *Heredity and Your Family's Health* (Baltimore: Johns Hopkins University Press, 1992), 58.

43. For example, in 1954, African American and white youth unemployment in the United States was equal, with African Americans actually having a slightly higher rate of employment in the age group sixteen to nineteen. By 1984, the African American unemployment rate had nearly quadrupled, while the white rate had increased only marginally.

44. Troy Duster, "Genetics, Race, and Crime: Recurring Seduction to a False Precision," in *DNA on Trial: Genetic Identification and Criminal Justice*, ed. Paul Billings (Plainview, N.Y.: Cold Spring Harbor Laboratory Press, 1992).

45. Eve K. Nichols, *Human Gene Therapy* (Cambridge: Harvard University Press, 1988).

46. Pa—92-03, Catalog of Federal Domestic Assistance 93.242 under authority of Section 301 of the Public Health Service Act, PL 78.410, 42 U.S.C. 241.

47. Grant no. R29 MH46925.

48. ROI MH38765.

49. James Q. Wilson, "To Prevent Riots, Reduce Black Crime," *San Francisco Examiner*, May 13, 1992. Reprinted from the *Wall Street Journal.*

50. Ibid. This scientizing of social stratification according to genetics continued in the best-selling book *The Bell Curve*, by Richard J. Herrnstein and Charles Murray (New York: The Free Press, 1994).

51. James Q. Wilson, "Point of View," *Chronicle of Higher Education*, June 10, 1992, A40.

52. Ibid.

53. Transcript of February 11, 1992, meeting of the National Mental Health Advisory Council.

54. More than double that of the previous decade.

55. Crime, mental illness, intelligence, and alcoholism.

56. K. S. Kendler and A. M. Gruenberg, "An Independent Analysis of the Danish Adoption Study of Schizophrenia, VI: The Relationship Between Psychiatric Disorders as Defined by the DSM-III in the Relatives of Adoptees," *Archives of General Psychiatry* 41 (1984): 555–64; K. S. Kendler and C. Dennis Robinette, "Schizophrenia in the National Academy of Sciences—National Research Council Twin Registry: A 16-Year Update" *American Journal of Psychiatry* 140, no. 12 (December 1983): 1551–63; Seymour S. Kety, "Genetic Aspects of Schizophrenia," *Psychiatric Annals* 6 (1976): 6–15; S. Kety, D. Rosenthal, P. Wender, and F. Schulsinger, "An Epidemiological-Clinical Twin Study on Schizophrenia," in *The Transmission of Schizophrenia*, ed. D. Rosenthal and S. Kety (Oxford: Pergamon Press, 1968), 49–63.

57. Glenn Garelik, "Born Bad? New Research Points to a Biological Role in Criminality," *American Health*, November 1993, 66–71.

[11]

Genetic Privacy: No Deal for the Poor

Karen Lebacqz

THE HUMAN GENOME PROJECT (HGP) promises to deliver genetic information that can be used for diagnostic or screening purposes long before cures or therapies for most conditions are available. Having knowledge of disease or of predisposition to disease may in some cases allow appropriate intervention to create an environment that ameliorates the effects of disease. Having knowledge without therapy raises troubling questions, however. Knowing that people are at risk for disease opens the possibility that they might be denied access to insurance, jobs, or other social goods such as adoption services. The perspectives of persons with disabilities, of members of minority groups, of women, and of other oppressed peoples suggest that one of the most troubling issues emerging from the HGP is the possibility of genetic discrimination. The research of Paul Billings and his colleagues indicates that genetic discrimination is already a reality.[1]

"What we want," says Ted Peters, "is information without discrimination."[2] The question is, can we get it? African American lawyer Patricia King warns that historically, genetic information has been used to reinforce negative stereotypes about racial groups and hence to support discrimination.[3] Information and discrimination have been linked. What makes us think that we can have information today without discrimination?

Our proposed answer is: privacy.[4] Since the heyday of the eugenics movement, we have established a strong tradition of making certain decisions "private" or personal and not easily subject to government regulation. Information that is private and not accessible to others or not legally allowed to be the basis for legislation cannot be used to discriminate against us. By making genetic information private, we will keep it from getting into the wrong hands, and it cannot be used to discriminate. So goes the argument.

I disagree. First, from a purely practical viewpoint, it is not clear that genetic privacy would be technically feasible. But more important, from a theological viewpoint, justice requires that social programs be judged from the perspective of the poor or oppressed. From the perspective of the oppressed, privacy is not an answer to the problem of genetic discrimination. Genetic privacy will not protect the poor and oppressed from discrimination. Privacy fails as a protection from discrimination not only because it may not be technically feasible, but also because it is politically and socially not feasible. Finally, the notion of genetic privacy may even be socially dangerous and run the risk of contributing to the very discrimination that we seek to avoid. After defining privacy and spelling out a theological perspective on justice, I will deal with the technical and social limits of privacy and then with its particular dangers for the oppressed.

What Is Privacy?

Nancy Wexler, long a pioneer in searching for a genetic test for Huntington's disease, is known to be at 50 percent risk of carrying the gene. If she carries the gene, she will get the disease—a devastating deterioration that usually begins in midlife and leads to severe incapacitation and death. During the Senate hearings on the Human Genome Project in 1989, then Senator Al Gore asked Wexler whether she had been tested for the Huntington's gene. Wexler replied, "These are very personal decisions."[5] While desiring to have a diagnostic test available, Wexler indicated that deciding to use it is a different story. If she were tested and found to carry the gene for Huntington's, then every time she spilled a glass of milk she would wonder whether it was just an ordinary accident or whether the onset of Huntington's disease was beginning. Wexler was not sure she wanted to live with that anxiety.

Wexler's story highlights three aspects of privacy as it is understood in the contemporary context in the United States. First, "privacy" means that I have the right not to know my genetic status if I choose not to. Although Wexler has devoted much of her professional life to developing an accurate diagnostic tool for Huntington's disease, she may nonetheless choose not to use that tool. This choice would be honored under her right of "privacy" of genetic information. She can choose not to know her genetic status.

Second, "privacy" means that if I do know my genetic status, I do not need to share that information with others. If Wexler does know her status, she is not telling others. Her refusal to tell has been honored.

A third "right" is also generally encompassed under the presumed right of genetic privacy: the right to use information to make decisions based on one's own values and without coercion from others. In the case of diseases for which prenatal diagnosis is a possibility, for example, a right of genetic privacy would mean that the parents of an affected fetus could choose whether or not to abort that fetus, based on their values and life choices.

Genetic privacy would therefore encompass three notions: a right not to know one's genetic status if one so chose, a right not to tell others or to have others know, and a right to use genetic information to make decisions within the framework of one's values. As we shall see, all three of these notions get challenged by the social, cultural, and practical realities in which we live. These challenges have particular implications for the oppressed, viewed from a theological perspective on justice.

A Perspective on Justice

Justice is a vast and unwieldy topic. Its requirements are as elusive as its claims are legion. Nonetheless, there is within the Jewish and Christian traditions a common core of understanding of the demands of justice that will not go away. This core provides grounds for a theological view of justice.

It begins in the conviction that the human community lives in response to God. Human justice, therefore, is a response to the justice of God. We seek justice because we understand the God who creates us to be a God of justice, and we seek the kind of justice that God desires and intends. "There is absolutely no concept in the Old Testament," writes Gerhard von Rad, "with so central a significance for all the relationships of human life as that of *sedaqah* [justice/righteousness]."[6] What von Rad posits of the Hebrew Scriptures could also be said of the Christian Scriptures: if justice is right relationship or the restoration of God's righteousness, then it is the central demand to which both Hebrew and Christian Scriptures point.

But what is right relationship? "In biblical thought, the litmus test for any society consists in the way it treats widows, orphans, alien farm workers—in short, the poor—in its midst."[7] The notion that God has a special concern for the poor is consistent through Jewish, Catholic, and Protestant perspectives.[8] Early church fathers, for example, understood that "charity" to the poor was really justice; it was returning to them what was their *due*.[9] In short, the poor or oppressed, the marginal groups, "become the scale on which the justice of the whole society is weighed."[10]

Further, justice is not simply the redistribution of goods. The poor have a different perspective on reality and will see it differently from the way the rich do. As the great American theologian Reinhold Niebuhr put it long ago, "Those who benefit from social injustice are naturally less capable of understanding its real character than those who suffer from it."[11] If justice is the restoration of right relationship, the rectification of injustice, then there must be an "epistemological privilege" of the poor, for only they can judge the true character of injustice and hence the demands of justice.[12] In assessing privacy as a solution to the problem of discrimination, for example, a biblical view would require that we ask what privacy looks like from the perspective of the poor.

Who are the poor, then? A number of terms are used in the Scriptures. *Anaw* means those who are humble before God; *dal* is the one who is weak; *ebyon* is a beggar; and so on.[13] These terms are both literal and metaphorical. They refer, on the one hand, to those who are economically poor. But they also connote all those who are without power, who are marginalized or oppressed. Thus, in contemporary society the "poor" would include members of minority groups; white women; persons with disabilities; gay, lesbian, and bisexual people; and anyone who is either economically disadvantaged or politically and socially disenfranchised. Their views on privacy become most crucial.

We turn, then, to see why privacy fails as a solution to the problem of discrimination, judged particularly from the perspective of the oppressed in our society.

Technical Problems with Privacy

At first glance, genetic privacy seems a sensible solution to the problem of discrimination, particularly in the U.S. context. We have a strong and growing tradition of protecting a right of "privacy" under the "penumbra" of the Bill of Rights.[14] Laws could be passed placing genetic information under a person's right of privacy. This would make it difficult for others to gain access to that information, and thus might protect against discrimination.

But there are good reasons to think that such a solution will not work. The first problems are technical and have to do with practical aspects of trying to keep genetic information private.

Information about genes will likely be garnered in a medical or medical research setting. In *Privacy for Sale: How Computerization Has Made Everyone's Private Life an Open Secret*, Jeffrey Rothfeder argues that it is extremely easy to gain access to medical records.[15] At present, no federal laws mandate limits to revelation of medical records. The Privacy Act of 1974 governs the privacy of medical records in the federal government, but this applies to only about 5 percent of medical databanks.[16] In a typical hospital, at least eighty people have access to medical records. If genetic information is included as part of a person's medical record, a wide range of people will already have access to this information.

Could specific laws be passed to ensure that genetic information is kept secret, even if other medical information is not? The social unfeasibility of this solution will be addressed below. Here the concern is whether this is technically feasible. With increasing computerization of medical records, electronic protections *could* be placed on certain information in a medical record. At present, however, when medical records are solicited, they are sent en masse.[17] No hospital has the staff to sort through medical records and pull out certain information to hold back. Everything that is in the record goes. An insurance company that has asked for information about a patient's asthma, for example, sees the entire medical record of that patient, not simply the portion pertaining to asthma. Under these practices, it is not feasible to say that genetic information can be part of a medical record and yet be kept secret. Anyone with legitimate access to the medical record will have access to the genetic information.

Who has legitimate access to medical records? The American Medical Association (AMA) has identified twelve categories of record seekers *outside* the health care establishment who routinely gain access to health care records. They include employers, insurers, government agencies, educational institutions, credit bureaus, and the media.[18] In many instances, the agency seeking the medical record has legitimate reason to do so—for example, to gain information to determine whether a prospective employee will be at risk, or whether a life insurance policy can be issued at a particular rate. So long as it is social policy to have private medical insurance, for example, it is legitimate for the prospective insurer to gain access to this information.

While consent may be needed for release of medical records in many states, the patient really has no choice but to consent if she or he wishes to be insured or to gain access to other social goods such as education. Since there is often no expiration date on the consent form, once consent is given, the company can seek access again and again using the same consent.[19] Thus, the information in a medical record—including genetic information if it is there—can determine whether a person is hired, fired, or insured, secures a business license, and performs many other social functions. "Medical-record information is now the key to many societal gatekeeping functions."[20]

Most patients probably assume that only organizations to which they give consent will have access to their records. However, that is not so. In one of the most famous cases, the Factual Service Bureau of Chicago "openly advertised its ability to procure confidential information about an individual without his authorization."[21] According to Dale Tooley, then district attorney in Denver, Colorado, the Factual Service Bureau was "ninety-nine percent successful in being able to pierce the medical records protection system of health care providers."[22] Although the report of the Privacy Protection Study Commission led to some concrete recommendations to eliminate some unauthorized access to health care records, Rothfeder claims that a Senate subcommittee investigating data selling in the 1980s found surreptitious trafficking in medical records "common and nationwide."[23]

In fact, Rothfeder offers several examples of well-known political figures whose political careers have been ruined when an opponent obtained their medical records and published them, either accurately or erroneously. Tommy Robinson, former U.S. representative, lost his bid for the governorship of Arkansas when his opponent leaked information from an inaccurate medical report that indicated he drank too much.[24] Rothfeder found it easy to obtain private information about then Vice President Dan Quayle and several other prominent figures. He implies that anyone with a little ingenuity can gain access to others' medical records.

However, the advent of computerization really makes dependence on the privacy of medical records unrealistic, in his view. When the Privacy Protection Study Commission did its report, the Medical Information Bureau, the largest company set up to facilitate the exchange of medical record information among insurers, was sending out roughly 3.5 million reports a year.[25] It now distributes more than 30 million reports on medical records each year to health, home, and car insurers.[26] Selling medical record information is big business: when restrictive laws were enacted in California, the Department of Motor Vehicles had to raise its fees to cover the revenue losses from the millions of reports it had been selling each year.[27] Computerization accounts for some of the increase in volume of the business.

Computerization can also make access easier. An extreme example of the ease of access by computer is that of Physicians Computer Network in New Jersey. This company offers physicians free computers, provided they are always linked to an electronic network so that data can be accessed at all times. The company claims that it gathers no personally identified data, but uses only aggregate data to sell to pharmaceutical manufacturers and others; nonetheless, it has constant access to the medical records of all those doctors' patients. This amounts to "an unregulated electronic pathway to personal information," says Rothfeder.[28]

Moreover, computerization of medical information can make it difficult to expunge inaccurate medical records. "David Castle" (a pseudonym) tried to get disability coverage while working out of his home. After three companies turned him down, he became suspicious. He discovered that the Medical Information Bureau was incorrectly releasing information that said he was HIV positive. The information was incorrectly recorded by a radiologist during a CT scan.[29] Once the information was in Castle's file, he could not get it expunged, but could get only a notation of error.

In another instance, "Charles Zimmerman" was turned down repeatedly for insurance. He finally discovered that because he had once attended an Alcoholics Anonymous (AA) meeting, his record at the Medical Information Bureau had him listed as "alcoholic," even though he had never been a drinker and had gone to AA to quit smoking. Computerized records, however, require that all data be reduced to codes; AA attendance is coded as alcoholism.[30]

Thus, there are numerous technical difficulties with depending on privacy to prevent discrimination. If genetic records are included with medical records, they will for all intents and purposes not be very private. Medical records are readily available, both with and without consent. They are used widely as gatekeeping functions in our society. Accurate or inaccurate, medical information about us is widely circulated. For all these reasons, depending on the privacy of genetic information or of medical records may not be technically feasible.

Social Constraints on Privacy

Depending on privacy as a defense against discrimination may also not be socially feasible. Precisely because of the link between medical information and certain gatekeeping functions, the privacy of medical information cannot be socially assured.

I noted above that one of the rights encompassed within the notion of genetic privacy is the right not to know one's genetic status if one chooses. Is a right not to know a socially feasible right? Evert van Leeuwen and Cees Hertogh of the Netherlands have argued strongly for a "right not to know" one's genetic information.[31] Respect for autonomy and choice, they suggest, implies that the genetic counselee has a right not to know. But this right is not absolute. While van Leeuwen and Hertogh resist overriding this right for societal economic reasons, they advance the idea that the right does not extend where others such as family members or offspring might be at risk. They might argue, for example, that an adult at risk for Huntington's *must* be tested if there are children in the family or if the person contemplates having children. Even those who argue strongly for a right not to know do set some limits on that right.

Van Leeuwen and Hertogh further observe that the right not to know depends on a certain infrastructure. Without full information, costly choices can be made: adults who choose not to know their genetic status may bear children with serious anomalies who need special supports and require considerable societal expenditures. These costs are currently mediated through an insurance system that aggregates risks. This is workable at present; however, van Leeuwen and Hertogh suggest that in the long run, the situation may change. Freedom and autonomy have their counterpart in social obligations. "Therefore, some expect that the emphasis will move away from respect for autonomy toward social obligations and liabilities, as advances continue to be made in genetic screening and counseling."[32] This could mean increased pressure on those who are at risk to "know" and also to use that knowledge to reduce economic burdens to the aggregate. Both the right not to know and the right to use knowledge in accord with one's values could be threatened.

Patricia King picked up precisely this risk.[33] The choices made by some women not to abort an HIV positive fetus have already led some commentators to propose that counseling should be more "directive," she states.[34] The presumed privacy of genetic counseling depends on the presumption that everyone will want to abort an affected fetus. It assumes certain values built into the system. Genetic differences have an impact on social relationships. The right not to know one's genetic status and the right to use knowledge in accord with one's own values will be limited by the range of gatekeeping functions to which medical status is linked. If genes are found to predispose for diseases that put us at risk, there will be social pressures to test for those genes and to use the information to reduce both risks and costs. This affects not only the right not to know one's own status, but also the right to keep that information from others and to use information in accord with one's values. It thus affects all three aspects of privacy as we currently understand it.

According to Paul Billings and Jonathan Beckwith, a recent Office of Technology Assessment study indicates that at least 10 percent of employers currently use genetic screening.[35] A representative of the biotechnology firm Oncor estimates that in the near future, genetic predispositions will be used as a risk factor for employment. Genetic screening is already widely used in some industries, and there will be pressures from industry and from the biotechnology firms that develop diagnostic tools to do additional screening.

Under current social practices, this makes eminent sense.[36] So long as access to health care is provided through insurance and employers carry responsibility for providing insurance to their employees, it will be in their interests to screen out persons whose health care places burdens on the insurance risk pool and drives up costs. It is no surprise that in Billings's review of forty-one cases of genetic discrimination, thirty-nine of them dealt with access either to insurance or to employment.[37]

As Patricia King noted in the Senate hearings, insurance is based on risk pooling: "If in a given case, a loss were *certain* to occur, insurance could not be obtained."[38] Insurance companies refuse to cover or limit coverage on costs attributable to preexisting conditions. Since genetic status is a preexisting condition, it stands to reason that insurance companies would not want to cover conditions related to genetic predispositions. They will therefore want to have access to this information. Henry Greely puts it bluntly: "The consequences of the genetics revolution for individual health insurance are straightforward: people who are known to be at higher risk for genetic illnesses will be denied insurance or sold insurance that excludes the conditions most important to them."[39] Genetic status cannot be a solely "private" matter as long as the system is insurance based. Information will be needed and sought in order to ascertain risks and determine pools. These social pressures undermine privacy as a protection against discrimination.

Rothfeder offers in some detail one of the cases located by Billings and his colleagues. A couple could not get health insurance because they were at risk for having a child with a serious genetic disorder. They had aborted their first child, who would have had the disorder. They had then borne a child who did not have the disorder. They were willing to sign a paper saying that they would abort any other fetus who would have the disorder. Nonetheless, they could not get insurance.[40]

Under such circumstances, their genetic status cannot be private. Others have a legitimate right of access to that information and will require that the information be utilized in certain ways. The couple's choice is surely constrained, if not coerced. It is precisely this that leads van Leeuwen and Hertogh to suggest that "instead of liberating people and making them able to shape their lives according to their own philosophy and religion, medical knowledge . . . becomes a social force in the distribution of inequality."[41]

The final extension of genetics into the public domain is demonstrated by an exchange from the Senate hearings of 1989: "Senator Pressler: 'If one wanted to determine the parentage of welfare children or something of that sort in certain instances, how accurate is DNA to determine a father of a child?' Watson: 'Totally accurate.'"[42] The potential use of genetic testing to determine the parents of children on Aid to Families with Dependent Children (AFDC) demonstrates how permeable is the boundary of genetic privacy.[43] It demonstrates that privacy cannot be guaranteed where there are overriding social and economic concerns. Pressures are built into the system to know, to share that knowledge, and to use it in certain ways. While we as individuals might want our genetic status to be private, as a society we are not structured to allow that, nor are we willing to pay the price for it. Privacy as a defense against discrimination fails not only for technical reasons, then, but also for social reasons. Because health status serves a gatekeeping function, knowledge of health status and access to information about it cannot remain solely private but must enter the public domain.

Privacy as a Dangerous Move

The exchange above regarding AFDC parents also demonstrates that the use of genetic technologies as gatekeepers for societal functions evokes deep issues of discrimination, stereotyping, and social "improvement." This brings us to a focus on the meaning of genetic technologies for the poor or oppressed. The effort to privatize genetic status has potentially dangerous consequences, which become apparent when we consider the social construction of genetic disease and of medical interventions. What is at stake in the development of genetic technologies is not merely the social organization of services in which medical records, including genetic information, become important as gatekeepers. What is at stake is the social construction of problems as "personal" rather than "societal."

Troy Duster puts the point forcefully: "The new genetics may help take social problems and turn them into personal problems."[44] For example, since diagnostic tools will be developed before therapies are in place, one of the most important uses of new genetic technologies will be in prenatal counseling and selective abortion. Making genetic information private means that this information will be shared with individuals or couples in a counseling setting as they make decisions about abortion. It will appear, then, that couples are exercising private decisions.

In fact, these decisions are not merely private. One survey of abortion in Bombay found that out of eight thousand abortions following prenatal diagnosis, only three were of male fetuses.[45] All of these abortion decisions appear to be private, and yet a clear ethnically patterned variation in sex determination results.[46] The privatizing of genetics diverts attention from the social circumstances such as cultural valuation of male children that influence patterns of technology use.

The social circumstances that influence patterns of technology use are generally determined by dominant groups. In the United States, the setting of prenatal diagnosis is already structured in a way that fits middle-class needs but not necessarily the needs of poorer classes. The system embodies certain values, and not others. Much prenatal diagnosis is by physician referral, for example. Since poor women of color often lack access to prenatal care, they are less likely to be referred for prenatal diagnosis and hence less likely to "benefit" from new genetic tests. To date, prenatal diagnosis has been largely available to and used by white, middle-class, well-educated women.[47] From the perspective of the oppressed, commments King, "medical institutions that direct the distribution of benefits that potentially will flow from gene mapping efforts may not be established in ways that maximize potential benefit for minorities and the poor."[48]

Although we may claim that everyone has genetic privacy, that will be true for some groups, but not for others.

If minority women had access to prenatal diagnosis, would it help them? Not necessarily, suggests King. If counseling is based on dominant values and is not sensitive to cultural differences or variations, then counselors and clients may be speaking a different language, and the information given may not enhance autonomy. "Without changes in our system of health care financing and access to health care delivery, there is no reason to believe that members of socially disadvantaged groups will benefit from the information they receive," argues King.[49]

Women of different cultures may make different decisions. We have seen above that there will be social pressures to utilize genetic information in certain directions, thus undermining privacy in the sense of the ability to make decisions based on one's values. These pressures are multiplied for women from oppressed groups. Will women who do not choose to abort a defective fetus be forced to do so in order to fit dominant values? Lest we think this idea is far-fetched, King reminds us that coercive reproductive policies have been promulgated in the past and are currently proposed for Norplant.[50] Racism and ethnic stereotyping remain unfortunate realities of our life together. A 1990 National Opinion Research Center survey concluded that whites persist in holding negative and false images of African Americans and members of other minority groups—for example, seeing them as lazy and less intelligent than whites.[51] The persistence of such stereotypes, declares King, raises questions as to whether major social institutions can be trusted to operate fairly where people of color are concerned.

Depending on privacy may therefore contribute to discrimination. The privacy argument leaves the system operating "as is" and takes attention away from health care reform, from social and economic factors that impact health status such as inadequate housing and lack of employment, and from the social construction of values.[52] Precisely because the genetic issue has been "privatized" and seen as a matter of "personal" decision making in a counseling setting, the values underlying that setting go unexamined. These values—the values supporting medical services—do not always meet the needs of the poor, nor do they reflect the values of oppressed communities.

If justice requires that systems be judged from the perspective of the oppressed, then privacy can be a dangerous notion. What looks good from one perspective may not look good from the perspective of oppressed groups: "If you were to go into a black community . . . and to ask people about their attitudes toward doing some of this [genome] work, one would get very harsh expressions in some instances, expressions of great fear because it is in that community that we realize that some of these eugenic issues did not end in 1920."[53] From the perspective of oppressed people, genetic privacy is not only impractical but may also be dangerous. The effort to privatize genetic information has the consequence of turning attention away from important social decisions and value issues.[54]

From the perspective of minority group leaders, King argues, "it might be preferable to advocate a reordering of priorities in light of the potential dangers the Genome Project poses to the interests of their constituents."[55] During the Senate hearings, King noted, "There are values that go into judging what is important and what is not important, and those values are only as good as the diversity of the people that are doing the judging"[56].

King argues for allocating resources to ameliorate social, environmental, and economic conditions that are known to influence health status rather than focusing on genetic contributions to disease prevention. A similar suggestion is made by Duster, who states that the incidence of lung cancer among African Americans has soared. Since the human gene pool does not change that rapidly, he remarks, this soaring rate cannot be a simple function of genetics. To focus on genetic marker studies for lung cancer rather than, for example, on cigarette billboard spending in African American communities is to avoid crucial social issues. To focus on genetic marker studies is to privatize what should be a public issue. To privatize genetics is to blind ourselves to harsh social realities that constrict private lives.

Genetic privacy is therefore dangerous because it shifts the ground of discussion and the angle of vision in ways that do not benefit the poor. "The discourse is always in terms of the medical, health, and scientific benefits," emphasizes Duster, "skewing the grounds upon which an informed debate about other social, cultural, and political questions can arise."[57] When the discourse takes place in the language of privacy, it distorts the grounds upon which an informed debate about social gatekeeping functions of genetics can arise.

The Limits of Privacy

Discrimination is a danger whenever new knowledge is developed. Privacy is not the solution. It is not technically feasible to trust privacy to ensure that information gained from the Human Genome Project does not result in discrimination. Further, under present circumstances it is not socially feasible. Medical information has a gatekeeping function in our society. There will be pressures to use this information to reduce costs for a number of social services and business ventures such as insurance. This is built into the very structure of our social arrangements.

Moreover, trusting privacy as a protective scheme assumes that there is nothing wrong with the delivery system in which genetic information is gained and disseminated. It assumes that the values underlying the current system are not flawed. But in fact, as Billings and Beckwith put it, "the action of genes and the impact of genetic testing will be highly dependent on the environments in which they are operating."[58] The present environment, at least in the United States, is one of "pervasive racism, ethnic stereotyping, and economic disparities in both health status and access to health care services."[59] In this context, the HGP appears to those who are poor and oppressed to offer dangers of discrimination in the very way that it structures the public discourse around disease, disability, and difference. Privacy does not change that discourse but only reinforces it: social problems become privatized and turned back onto oppressed peoples. Troy Duster is correct to assert that "whether this new genetic knowledge is an advantage or a cross depends . . . upon where one is located in the social order."[60] Only a rearrangement of that order and attention to the values hidden in the system will prevent discrimination. Privacy may have its place, but it is not a protection against discrimination.

Notes

1. Paul R. Billings et al., "Discrimination as a Consequence of Genetic Testing," *American Journal of Human Genetics* 50 (1992): 476–82.

2. Ted Peters, "Genome Project Forces New Look at Ethics, Law," *Forum for Applied Research and Public Policy* 8, no. 3 (fall 1993): 7. See Ted Peters, *For the Love of Children: Genetic Technology and the Future of the Family* (Louisville: Westminster John Knox, 1996), 89.

3. Patricia King, "The Past as Prologue: Race, Class, and Gene Discrimination," in *Gene Mapping: Using Law and Ethics as Guides*, ed. George J. Annas and Sherman Elias (New York: Oxford University Press, 1992), 99.

4. The trend toward privacy is not the only difference between the contemporary social situation and the earlier times, of course. There is always a desire to believe that we are wiser now and will use information only for good, where our foreparents might have used it for evil. Such assumptions about progress in wisdom raise questions about the social construction of reality. One aspect of that construction will be addressed later on in this chapter.

5. U.S. Senate, Hearing before the Subcommittee on Science, Technology and Space of the Committee on Commerce, Science, and Transportation, 101st Congress, November 9, 1989, 105 (hereinafter: Senate Hearing 1989).

6. Gerhard von Rad, *Old Testament Theology*, vol. 1 (New York: Harper Bros., 1962), 370.

7. John A. Coleman, "A New Catholic Voice on Social Issues," in *A Cry for Justice: The Churches and Synagogues Speak*, ed. Robert McAfee Brown and Sidney Thompson Brown (New York: Paulist Press, 1989), 64.

8. See, e.g., John R. Donahue, "Biblical Perspectives on Justice," in John C. Haughey, ed., *The Faith That Does Justice* (New York: Paulist Press, 1977); David Beale, "Jewish Statements on Social Justice," in R. M. Brown and S. T. Brown, eds., *A Cry for Justice* (New York: Paulist Press, 1989); the United Presbyterian Church, U.S.A., "Christian Faith and Economic Justice," pp. 364–99 in Minutes: 196th General Assembly, P.C.U.S.A., May 29–June 6, 1984 (New York: Office of the General Assembly).

9. See Charles Avila, *Ownership: Early Christian Teaching* (Maryknoll, N.Y.: Orbis Press, 1983).

10. Donahue, "Biblical Perspectives on Justice," 78.

11. Reinhold Niebuhr, *Moral Man and Immoral Society* (New York: Charles Scribner's Sons, 1932), 80.

12. See Karen Lebacqz, *Justice in an Unjust World* (Minneapolis: Augsburg, 1987), 61.

13. See Conrad Boerma, *Rich Man, Poor Man—and the Bible* (London: SCM Press, 1979).

14. The Supreme Court first enunciated this right in *Griswald v. Connecticut* in 1965; it has become entrenched in reproductive decision making since the 1973 *Roe v. Wade* and *Doe v. Bolton* decisions on abortion.

15. Jeffrey Rothfeder, *Privacy for Sale: How Computerization Has Made Everyone's Private Life an Open Secret* (New York: Simon & Schuster, 1992), chap. 8. See the cover story exposé by Joshua Quittner, "Invasion of Privacy," *Time*, August 26, 1977, 28–35.

16. John Eckhouse, "Why Nothing's Secret Anymore," *San Francisco Chronicle*, March 15, 1993, E6.

17. Rothfeder, *Privacy for Sale*, 181.

18. Ibid., 180.

19. Ibid., 177. The Privacy Protection Study Commission concluded that consent forms were often so broadly worded that the patient, "in effect, signs away all control over what is disclosed and what may be done with it thereafter." See *Personal Privacy in an Information Society* (Washington, D.C.: Government Printing Office, 1977), 286.

20. Privacy Protection Study Commission, *Personal Privacy*, 281.

21. Ibid., 174.

22. Quoted in Rothfeder, *Privacy for Sale*, 182.

23. Ibid.

24. Ibid., 175–76.

25. Privacy Protection Study Commission, *Personal Privacy*, 160.

26. Rothfeder, *Privacy for Sale*, 187.

27. Eckhouse, "Why Nothing's Secret Anymore," E6.

28. Rothfeder, *Privacy for Sale*, 193.

29. Ibid., 184.

30. Ibid., 185–86.

31. Evert van Leeuwen and Cees Hertogh, "The Right to Genetic Information: Some Reflections on Dutch Developments," *Journal of Medicine and Philosophy* 17, no. 4 (August 1992): 382–91.

32. Ibid., 385.

33. King, "The Past as Prologue," 95–107.

34. Ibid., 105.

35. Paul Billings and Jon Beckwith, "Genetic Testing in the Workplace: A View from the U.S.A.," *Trends in Genetics* 8, no. 6 (June 1992): 200.

36. See King, "The Past as Prologue"; also David A. Peters, "Risk Classification, Genetic Testing, and Health Care: A Conflict Between Libertarian and Egalitarian Values?" in this volume.

37. Billings et al., "Discrimination as a Consequence of Genetic Testing," 478. Billings and his colleagues define genetic discrimination as discrimination against an individual or members of that person's family "solely because of real or perceived differences from the 'normal' genome" (477).

38. Senate Hearing 1989, 82, emphasis added.

39. Henry T. Greely, "Health Insurance, Employment Discrimination, and the Genetics Revolution," in *The Code of Codes: Scientific and Social Issues in the Human Genome Project*, ed. Daniel J. Kevles and Leroy Hood (Cambridge: Harvard University Press, 1992), 266.

40. Rothfeder, *Privacy for Sale*, 188–90.

41. Van Leeuwen and Hertogh, "The Right to Genetic Information," 389.

42. Senate Hearing 1989, 51.

43. It also demonstrates that the entanglement of new genetic technologies into social policies is written into the development of the genome project from its inception.

44. Troy Duster, *Backdoor to Eugenics* (New York: Routledge, 1990), 74.

45. Ibid., 33.

46. Ibid., 35. 47. King, "The Past as Prologue," 103.

48. Ibid., 100.

49. Ibid., 106.

50. Ibid. See also Marque-Luisa Miringoff, *The Social Costs of Genetic Welfare* (New Brunswick, N.J.: Rutgers University Press, 1991), 13: "Individuals are increasingly seen as having a right to genetic health: parents have a *duty* to ensure the outcome. The technological ability to accomplish these outcome imposes a new imperative" (italics in original).

51. King, "The Past as Prologue," 97.

52. We often assume, for example, that medical interventions are benefits. However, portions of the Deaf community resist efforts to do surgery on children who are deaf to restore their hearing. Thus, medical intervention may not always be seen as a benefit. See Edward Dolnick, "Deafness as Culture," *Atlantic Monthly*, September 1993, 37–53.
53. King, Senate Hearing 1989, 107–8.
54. See also Miringoff, *The Social Costs*, chap. 1.
55. King, "The Past as Prologue," 107.
56. King, Senate Hearing 1989, 107.
57. Duster, *Backdoor to Eugenics*, 129.
58. Billings and Beckwith, "Genetic Testing in the Workplace," 201.
59. King, "The Past as Prologue," 96.
60. Duster, *Backdoor to Eugenics*, 92.

Index

Abelson, Philip H., 103n. 10
ABO. *See* blood types
Aborigines, 221
abortion, 6–8, 127, 132, 140, 152, 153, 154, 192, 248
Abram, Morris, 136
actuarial justice, 207, 210
Adam and Eve, 23, 116, 145
adenine, 51
African Americans, 20, 79, 82, 188, 196, 229, 249, 250
Akiva, Rabbi, 191
Alamoudi, Abdurahman, 12
Albrecht, Paul, 123
Alcohol, Drug Abuse, Mental Health Administration (ADAMHA), 232
Alcoholics Anonymous (AA), 245
alcoholism, 18, 62, 118
alienation, 115
alleles, 54
Alves, Rubem, 126, 127
Alzheimer's disease, 3, 6, 60, 62
American Medical Association, 243
American Muslim Council, 12
amino acids, 50
amniocentesis, 154, 160, 162
Anderson, W. French, 26
Andre, Richard, 184

antenatal testing. *See* embryos, routine testing of
Aquinas, Thomas, 145–46
Aristotle, 122, 145
arteriosclerosis, 6
artificial insemination, 125, 140, 148, 158
Aryans, 183, 186, 189
Asimov, Isaac, 168
atherosclerosis, 4
attention deficit hyperactive disorder (ADHD), 62
Augustine, Saint, 23, 116, 145
autonomy, 75, 168, 211–12
autosomes, 53
awe, 29

bacteria, 58
Baker v. State Bar of California, 18
base pairs (bp), 51, 57, 60, 65
basic research, 84, 91, 99
Beckwith, Jonathan, 246, 251
bees, 115
beneficence, principle of, 26
big bang theory, 24
big science, 34n. 2, 65
Billings, Paul, 239, 246, 251
Binet test, 186